ECOLOGICAL FORM

Ecological Form

System and Aesthetics
in the Age of Empire

Nathan K. Hensley
and Philip Steer

Editors

FORDHAM UNIVERSITY PRESS
New York 2019

Copyright © 2019 Fordham University Press

All rights reserved. No part of this publication may be reproduced, stored in a retrieval system, or transmitted in any form or by any means—electronic, mechanical, photocopy, recording, or any other—except for brief quotations in printed reviews, without the prior permission of the publisher.

Fordham University Press has no responsibility for the persistence or accuracy of URLs for external or third-party Internet websites referred to in this publication and does not guarantee that any content on such websites is, or will remain, accurate or appropriate.

Fordham University Press also publishes its books in a variety of electronic formats. Some content that appears in print may not be available in electronic books.

Visit us online at www.fordhampress.com.

Library of Congress Cataloging-in-Publication Data available online at https://catalog.loc.gov.

Printed in the United States of America

21 20 19   5 4 3 2 1

First edition

CONTENTS

Introduction: Ecological Formalism; or, Love among the Ruins
NATHAN K. HENSLEY AND PHILIP STEER   1

Part I METHOD

1. Drama, Ecology, and the Ground of Empire: The Play of Indigo
SUKANYA BANERJEE   21

2. Mourning Species: *In Memoriam* in an Age of Extinction
JESSE OAK TAYLOR   42

3. Signatures of the Carboniferous: The Literary Forms of Coal
NATHAN K. HENSLEY AND PHILIP STEER   63

Part II FORM

4. Fixed Capital and the Flow: Water Power, Steam Power, and *The Mill on the Floss*
ELIZABETH CAROLYN MILLER   85

5. "Form Against Force": Sustainability and Organicism in the Work of John Ruskin
DEANNA K. KREISEL   101

6. Mapping the "Invisible Region, Far Away" in *Dombey and Son*
ADAM GRENER   121

Part III SCALE

7. How We Might Live: Utopian Ecology in William Morris and Samuel Butler
BENJAMIN MORGAN   139

8. From Specimen to System: Botanical Scale
and the Environmental Sublime in Joseph Dalton
Hooker's Himalayas
   LYNN VOSKUIL   161
9. "Infinitesimal Lives": Thomas Hardy's Scale Effects
   AARON ROSENBERG   182

Part IV   FUTURES

10. Electric Dialectics: Delany's Atlantic Materialism
    MONIQUE ALLEWAERT   203
11. Satire's Ecology
    TERESA SHEWRY   223

Afterword: They Would Have Ended by Burning
Their Own Globe
   KAREN PINKUS   241

Acknowledgments   249
List of Contributors   251
Index   253

Ecological Form

INTRODUCTION

Ecological Formalism; or, Love Among the Ruins

Nathan K. Hensley and Philip Steer

> Subjection of Nature's forces to man, machinery, application of chemistry to industry and agriculture, steam-navigation, railways, electric telegraphs, clearing of whole continents for cultivation, canalisation of rivers, whole populations conjured out of the ground—what earlier century had even a presentiment that such productive forces slumbered in the lap of social labour?
>
> —KARL MARX AND FRIEDRICH ENGELS, *The Communist Manifesto* (1848)

> In the gloom she did not mind speaking freely.
>
> —THOMAS HARDY, *Tess of the D'Urbervilles* (1891)

For Marx and Engels in 1848, European modernity was a world-demolishing juggernaut, an engine of vast productivity and vaster catastrophe. To these most sensitive observers of contemporary life, the new industrial age, powered by burned coal and the brute labor of newly urbanized masses, was most recognizable as a terraforming project. Altered chemistry, moved earth, rerouted rivers: Capitalism was a continent-clearing attack on nature at world scale, a magic act by which plants, wealth, and even human populations could be created as though from nothing—"conjured out of the ground." In this steam-driven and electrified present, humankind or an empowered subset of it, enriched by extraction and aided by machine technology, could enslave the very forces of nature (*Naturkräfte*), and, like Xerxes whipping the Hellespont in Herodotus's famous parable of outlandish pride, alter the flow of waters on earth. Modernity's self-inflicted demise was incipient or imminent to Marx and Engels: They anticipated the bourgeois world's terminal crisis as future revolution, augured in stories of chastened hubris and tragedy inherited from the Greeks.

To twenty-first-century observers, by contrast, the generalized death drive of western life is palpable, legible, here and now. The earth and its interlocked systems now seem a material laboratory for proving not just Marx's observation about capitalism's tendency toward suicide, but also Freud's late-career discovery, stunning even to himself, that a sentient organism might somehow desire, and then willingly pursue, its own destruction.[1] Ice shelves collapse and glaciers retreat; particulate plastic swirls in eddies the size of continents; species vanish at rates not seen since an asteroid restarted the clock of evolution; and the weather of our daily lives is a coded message that we have altered the world forever. But despite being locked into this "terminal crisis of the Holocene," we charge onward, unwilling or unable to replace the languages of growth, mastery, and progress we inherit from the era of Marx and Freud.[2] "It is painful to say," explains Jeremy Davies, "that efforts to keep climate change to even minimally tolerable levels may well be futile by now. . . . [T]he feedback mechanisms already triggered mean that no human power whatsoever can halt the changes that are now under way."[3] The world-enslaving omnipotence Marx and Engels ambiguously celebrated has transformed into its opposite, helplessness, and as though fulfilling Victorian prophecies modernity seems to have dug its own grave.

Ecological Form is about how we might think about the nineteenth century—about how we *need* to do so—as we come to terms with a damaged and seemingly diminished present. What can the Age of Coal tell us about the Age of Man? What messages might speak across the divide that separates the subjection (*Unterjochung*) Marx identified in London and Manchester from our own moment of catastrophic mastery? And in what ways does the legacy of extractive imperialism in the nineteenth century continue to shape experience now? In his crucial and early effort to think environmental and colonial histories together, Rob Nixon refers us to "the long dyings—the staggered and staggeringly discounted casualties, both human and ecological," that modernity leaves behind.[4] These dyings are the necessary aftereffects of an economic order that, by design, sees the nonhuman world as a theater for accumulation, dispossession, and capture: We could call it neoliberal ecocide. The past becomes new from the vantage of every present, and each age sees itself in what came before. But as Nixon points out, and as our own daily experience verifies, the disastrous modernity that so shocked Marx and Engels lives with us still.

Victorian England was both the world's first industrial society and its most powerful global empire: The nineteenth century therefore stands as the origin of not just the irreversible ecological degradation we have inherited from our nineteenth-century forebears, but also the global interconnection and vast asymmetries of power that are the legacies of the British

Empire in the present. Given that the Victorian Empire's world-spanning configuration was the first political project in history to be powered almost exclusively by fossilized plant life, it follows that the carbon-saturated atmosphere we breathe today is, in both metaphorical and brutely chemical senses, the atmosphere of the British Empire.

The fact that we inhabit this extended carbon modernity makes impossible any simple attempt to cleave then from now, them from us. The increasingly lethal pH levels of world oceans, for example, which now bleach to death the coral reefs that in 1842 charged Charles Darwin with an almost erotic excitement, are rising because ocean water—operating at timescales only unevenly synchronized with the other human and earth biorhythms to which it is linked—continues to assimilate CO_2 from fuel burned since the days when chimneys choked the residents of Manchester.[5] These same seas now rise to drown out precarious populations of subsistence farmers and fishermen in places like, say, the floodplains of Bangladesh, a Muslim-majority nation born in the catastrophic 1947 Partition of Bengal. Such thoroughly modern crises sit at the conjuncture of demography, political economy, and climate change, and have as their condition of possibility the geopolitical and demographic carving-up accomplished by the British Empire. The uncanny but perversely material presence of the Victorian era's coal-fired and imperial past, then, means that our new contemporary is best viewed as but a moment in a much longer unfolding, a longer *durée* over which the nineteenth century looms like the angel in Walter Benjamin's famous essay, in whose eyes history becomes not a series of discrete events but "one single catastrophe which keeps piling wreckage upon wreckage and hurls it in front of his feet."[6] Resilience is part of this story, too. But our geophysical and demographic links to the Victorian moment mean that human and nonhuman scenes of subjection must be imagined together at this longer, even geological scale.

"The Anthropocene has reversed the temporal order of modernity," writes the novelist Amitav Ghosh. That is because "those at the margins are now the first to experience the future that awaits us all."[7] One need not so readily adopt the bleak confidence of Ghosh's assessment to see that our anthropogenic present has scrambled the narrative templates and historical logics previously available for organizing experience. Rather than reversing modernity's order, our only lately dawning awareness of climate change might be said to have thrust the very premise of modernization—like its corollaries, dear to Marx no less than to his liberal enemies, progress and freedom—into crisis. This crisis pushes us to "the limits of historical understanding," in Ghosh's words, and exposes extant conceptual models as inadequate for construing our current conjuncture, never mind for thinking beyond it.[8] In such a situation, the task of criticism cannot be simply to switch our attention to environmental themes or ecological motifs and carry on otherwise as usual. The challenge is not about content but about form, not about accumulating more information but about reframing the methods by which we understand it. Ghosh himself describes his own previous resistance to incorporating into the plot-structure of his fiction the "unbearably intimate connections over vast gaps in time and space" that climate change generates. But "we are confronted suddenly," he notes, "with a new task: that of finding other ways in which to imagine the unthinkable beings and events of this era."[9] Under the pressure of our new climatological present, the very structure of thought must change.[10]

The eleven new essays commissioned for this collective project aim to show how one Anthropocene first emerged into visibility in the nineteenth century. Together these interventions aim to demonstrate the diligence and acuity with which certain Victorian writers experimented with new formal techniques, and generated new models for thinking, in order to comprehend the two massively networked and often violent global systems that organized their experience, and that, we suggest, continue to organize ours: the British Empire and the Industrial Revolution's carbon economy. The weblike networks of George Eliot's realism or Darwin's tangled banks are just two ways in which Victorian thinkers imagined mutual imbrication at planetary scale: Political economy, evolutionary biology, thermodynamics, early geology, and imperial administration were others. In these domains and more, the humanities continue to come to grips with the question of how the increasingly palpable fact of anthropogenic climate change will impact its own methods.[11] Nearly a decade after Dipesh Chakrabarty's groundbreaking essay listed four theses for a new Anthropocene method, Ian Baucom and Matthew Omelsky still find cause to ask: "What does it mean to generate knowledge in the age of climate change?"[12] *Ecological Form* engages the persistent challenge of climate change method by (1) contributing a historical account of the period most consequential in framing the horizons of contemporary earth systems and our relations to them, the nineteenth century,[13] and (2) by widening that problem of ecological thought to imperial, and therefore political, scale. Together, the authors gathered here demonstrate the need to rethink the procedures of cultural analysis in light of the fact that the Age of Coal, the Age of Empire, and the Age of Man are one and the same.

Victorian Studies is well positioned to speak on the topic of our climatological disaster. As a field, it has generated a set of path-breaking works that have helped us see the nonhuman environment as central to the production of culture in modernity. Jesse Oak Taylor's *The Sky of Our Manufacture: The London Fog in British Fiction from Dickens to Woolf* (2016) and Allen MacDuffie's *Victorian Literature, Energy, and the Ecological Imagination* (2014) developed canonical statements by Gillian Beer and others to resensitize critics to the Victorians' incipient ecological thinking. Devin Griffiths's *The Age of Analogy: Science and Literature Between the Darwins* (2017) has shown how knowledge generated in botany and evolutionary science came to shape historicist and literary method. Other monographs, by Justine Pizzo, Tobias Menely, and several others gathered in this volume, are now in process, and a volume entitled *Anthropocene Reading: Literary History in Geologic Times* (2017), edited by Taylor and Menely, has recently drawn on its editors' expertise in nineteenth-century archives to situate the Anthropocene "as a geohistorical event that may unsettle our inherited practices of reading."[14] In addition to this robust and growing conversation about ecology and the field's longstanding engagement with questions of race, violence, and empire, Victorian Studies has also been at the forefront of a renewed attention to literary form and its relationship to social and political structures. From signal early works such as Franco Moretti's *An Atlas of the European Novel, 1800–1900* (1998), to Caroline Levine's more recent *Forms: Whole, Rhythm, Hierarchy, Network* (2014), the nineteenth century has been the testing ground for new and experimental accounts of the cultural work accomplished by narrative and poetic structure.

This book aims to bridge and expand these too-often discrete conversations by setting into motion what we call *ecological formalism*: an approach that reconsiders Victorian literary structures in light of emergent and ongoing environmental catastrophe; coordinates these "natural" questions with social ones; and underscores the category of form—as built structure, internal organizing logic, and generic code—as a means for producing environmental and therefore political knowledge. *Ecological Form* argues that the resources of ecological thinking can enable Victorian Studies to bridge the false divide between environmental history and the criticism of empire. This divergence between "natural" and social concerns was symptomatically expressed in the near-simultaneous publication of two books aspiring to define their subfield: historian Alfred Crosby's *Ecological Imperialism: The Biological Expansion of Europe, 900–1900* (1986) and Patrick Brantlinger's *Rule of Darkness: British Literature and Imperialism, 1830–1914* (1988). Where the first posited a biological account of empire, the other focused on culture, and neither touched the other's domain. With important exceptions, this schism between ecological and postcolonial approaches continues to play out as a split tradition, one concerned with nonhuman or "natural" actors, stories, and causal accounts and the other with human—that is, sociopolitical—ones.[15] Yet if climate change teaches us anything, it is that these stories must be told together. The essays in this volume bear out what is already known to the precarious human beings inhabiting modernity's sacrifice zones: Jason Moore's sense that sociopolitical dynamics and "natural" ones mutually inform one another, and what Jennifer Wenzel, reading Frantz Fanon, calls "the indivisibility of the social and the ecological."[16] *Ecological Form* coordinates a historically attuned focus on ecology with the sensitivity to human vulnerability long associated with the critique of imperialism. This enables us to show collaboratively how nineteenth-century culture developed powerful aesthetic and political tools for engaging with intractable problems that remain our own: problems of interconnection and asymmetry, distance and intimacy, system and disaster. This is why we now find ourselves thinking about the trains in *Tess of the D'Urbervilles*.

There are trains in *Tess*, after all—lots of them. Reading Victorian literature from within our great derangement presses us to notice the fossil fuel economy enciphered in those pages—and to look on as the everyday settings of realist novels like *Tess* transform under our new sensitivities into elaborate maps of the combustion, storage, and conversion of carbon-based fuel. This carbon infrastructure is a matter of simple referential content, yes: overt references, in *Tess*, to train rides and steam-powered harvesting machines. But the energy regime of coal also, and more importantly, conditions how the very form of this novel—and, we suggest, the Novel more broadly—can be organized. During the heady days when Angel is courting Tess in the Vale of Froom, the lovers drive one wet evening to deliver milk-cans to the nearest railway station. Tess witnesses the train being loaded and, "susceptible . . . [to] the few minutes of contact with the whirl of material progress," begins to wonder about the complex and impersonal connection that links her to a broader system of consumption and exchange:

> "Londoners will drink it at their breakfasts to-morrow, won't they?" she asked. "Strange people that we have never seen."
>
> "Yes—I suppose they will. Though not as we send it. When its strength has been lowered, so that it may not get up into their heads."
>
> "Noble men and noble women, ambassadors and centurions, ladies and tradeswomen, and babies who have never seen a cow."
>
> "Well, yes; perhaps; particularly centurions."

"Who don't know anything of us, and where it comes from; or think how we two drove miles across the moor to-night in the rain that it might reach 'em in time?"[17]

More than simply representing the incursion of modernity into the allegedly feudal space of these hinterlands, Hardy's train station shows the novel imagining its differential social geographies in systemic terms. Tess's time in the Vale of Froom is defined by the fecundity of nature, the private interiority of heterosexual attraction, and geographic isolation. But here, if just briefly, those scenes of pastoral bliss and "natural" unity are revealed to be connected to a wider national economy—even (with "centurions") an imperial one. In such details Hardy's novel discloses obliquely the mutually sustaining relationship between, on the one hand, a modernizing, urbanizing metropolitan society in which babies have never seen cows, and, on the other, the productivity and effulgence Hardy is at pains to link to a category called nature. It is what Ghosh called an unbearable intimacy. And as in all such intimacies, distinction begins to break down: If we pause a moment at this obscure provincial railway station, we begin to wonder how natural that pastoral landscape really is. Hardy describes the station's lamp as a "poor enough terrestrial star." This modern star is "in one sense of more importance to Talbothays Dairy and mankind than the celestial ones to which it stood in such humiliating contrast."[18] The dairy in the Vale of Froom—governed by the rhythms of the railway, lit by dingy stars, its very existence dependent on a metropolitan market for milk—begins to appear as inextricably linked to, and therefore a product of, the very carbon modernity the novel conscripts it symbolically to contrast.

The dialectical codependence of nature and culture modeled here is what Jason Moore has described as the operative dynamic of all value creation under capitalism. Anna Tsing introduces us to the inevitable collaborations and contaminations between these seemingly stable categories, while Derrida in 1966 generated an early and powerful form of his method by showing how the categories of nature and culture collapse, in Claude Lévi-Strauss, to indistinction.[19] A century earlier Darwin himself butted up against the shocking realization that humanity was also part of nature, and imported a Biblical idiom uneasily to patch over the fact that humankind itself might one day end up as just another loose branch on the tree of life. More telling for us, the collapsed division between nature and culture playing out in Hardy's countryside railroad station is also the structuring condition of the novel as such. It is, at least, if we are to believe Georg Lukács, who in *The Theory of the Novel* (1920) ascribed the advent of novel form itself to modernity's effort to come to terms with its relation to a lapsed and absent nature.[20] Yet more narrowly, the dynamic Hardy plays out at the level of symbol in the railway lamp is also the animating tension of Hardy's preferred figurative register *within* the novel form, pastoral. As Raymond Williams notes, this mode only comes into existence when an urbanizing modernity ("culture") began to require a poetic other ("nature") and, to fill that need, generated for its own delectation and self-affirmation images of the country as "an enamelled world," where labor is erased and social dynamism stilled into something like landscape.[21] Following orthodox materialist practice to focus on labor relations rather than the energy forms coproducing them, Williams pins this shift in the figuration of the country and the city to "the Industrial Revolution": a periodization that discloses how fully our entire range of aesthetic templates depends upon—is unthinkable without—a nascent and then maturing fossil economy.[22]

Growing up in the Vale of Blackmoor, apparently removed from that riotous modernity, Tess Durbeyfield seems to embody the local knowledge that the Victorian novel has taught us to expect from precapitalist life.[23] To her, Hardy's narrator observes, "[e]very contour of the surrounding hills was as personal . . . as that of her relatives' faces; but for what lay beyond her judgment was dependent on the teaching of the village school."[24] This appearance of geographic stasis primes the reader's expectation that Tess will, following the logic of *bildung*, soon transcend the limits to her individual growth. But the historical sweep of the novel instead makes clear that Tess's improvement is really a tale of decline, even tragedy, cast at evolutionary scale. Tess bears the corrupted name of a formerly powerful aristocratic family, whose bones lie interred and forgotten around the Wessex countryside, and she and her dispossessed family will ultimately spend a night encamped in one such graveyard, "their carvings . . . defaced and broken; their brasses torn from the matrices." Where the "spoliation" of her ancestral home reminds Tess "that her people were socially extinct," that last Darwinian term reminds *us* that it is not simply Tess's name but her very biotic existence, her "blood," that's been infected and determined by its evolutionary predecessors.[25] Her body is a holdover from a deep past over which her present self, only feebly able to act in the present, has no control at all. What form, this book asks, could map such unbearably intimate systems of entanglement? What cognitive tools might draw connections that reach not just between and among multiple bodies and landscapes—Wessex, the Arctic, Brazil—but across a timescale that links feudal crypts and Roman ruins with the biophysical histories, themselves accrued over eons, of the animals and plants thrown together in this fecund but doomed rural countryside?

The Victorians invented ecology: The term first entered English usage in *The Academy*, a British scientific journal, in 1875, and while the word had been coined in German in 1866, by Ernest Haeckel, Haeckel's "presentation of the term . . . embodies concepts that come straight from [Darwin's] *Origin of Species*."[26] Over the course of England's most modern century, the conceptual dilemmas of human beings' intertwinement with a world newly understood as "evolving, relational, and holistic" were felt most intensely as problems of intellectual scale.[27] How could the individual instance and the massive system be imagined at the same time? And how could any single actor within such a network envision resisting, or even altering that network? For many of the century's most sophisticated observers, these issues of scale were also problems of aesthetic form. By what figural means, these thinkers asked, could one hope to represent in a coherent literary or artistic work an entire ecosystem, where no single phenomenon can be abstracted from that system of mutual codependence? The still-startling caesura in the first third of George Eliot's *Middlemarch* (1872)—"but why always Dorothea?"—yanks us out of the focalizing, individualizing logic of novel form only to reassert a more capacious, multinodal version of that form, imagined through the techniques of sympathy.[28] But that is only a particularly gripping instance of the many means by which Victorian thinkers imagined systems and form together. As Eliot's example indicates, these aesthetic concerns in turn extend to the domain of conceptual or philosophical method. And if the disastrous entanglement between human and world in the era of coal-powered globalization generated dilemmas for literary and aesthetic presentation, those dilemmas do not go away when we, as later critics or readers, write and think about those (historical) problems. To the contrary, they become more acute, registering in, for example, our choice of intellectual objects; our delimitation of acceptable periods for analysis; the management our thinking and writing performs between instance and category, the particular and the general, the node and the system. *Ecological Form* addresses the vexed dilemmas of what Baucom and Omelsky call "knowledge in the age of climate change" by separating the problem into four domains—method, form, scale, and futures—which correspond to the book's sections.

The first section, on "Method," stages its arguments at the level of conceptual procedure to offer models for rethinking nineteenth-century studies through ecological form. These essays directly question how our objects of inquiry, preoccupations, and geographical horizons change in light of the new perspectives afforded by ecocritical theory, formal analysis, and critical studies of the Anthropocene. To ask what it means to acknowledge the fundamentally ecological nature of colonialism, Sukanya Banerjee focuses on the industrial cultivation of indigo in nineteenth-century Bengal and Bihar. This concern leads her to drama—specifically, to Dinabandhu Mitra's play, *Neel Darpan* (1860)—and to show how the "groundedness" of dramatic text and performance, rather than its transnational mobility, might enable us to conceive of the complex intersections between colonizer, colonized, and non-human agents. Another seemingly "grounded" form, the elegy, sits at the center of Jesse Oak Taylor's contribution: Alfred Tennyson's *In Memoriam* (1850) offers a test-case for an Anthropocene literary history because it allows us to revisit the Victorian archive with an awareness of our species' geological agency. In an age of mass extinction, to mourn a species, Taylor argues, is to freight that species with ethical and political consequence; to read *In Memoriam* as an elegy *for* the Anthropocene is therefore to grasp how shared loss might provide the basis for new forms of community and politics. Turning from the Anthropocene to the fossil fuels that have produced it, Nathan K. Hensley and Philip Steer take up the question of coal's paradoxical invisibility as an energy system in the Victorian novel. Omnipresent but strangely inapprehensible, the spectacular energy surplus of coal power finds form in narrative structures that give shape to, or seek to stall, the forcible opening of bounded societies to a global economy. Turning this "hermeneutics of coal" on Elizabeth Gaskell's *Cranford* (1853) and *North and South* (1855) as well as on Joseph Conrad's spiraling *Nostromo* (1904), Hensley and Steer also disclose the decisive but disavowed role coal plays in our most influential critical accounts of political reading, from Catherine Gallagher to Fredric Jameson.

The second section uses the category of "Form" to coordinate the dilemmas of environmental and political ecology described previously. These essays explore the capacity of Victorian forms not just to represent ecological and economic systems as content or theme, but to model them in their own organizational and imaginative structures. Elizabeth Carolyn Miller's chapter highlights the kinds of temporal awareness and economic thinking that arise when we think in terms of energy. For Miller, George Eliot's *The Mill on the Floss* (1860) is not only a female *bildungsroman* or modernity story but a sophisticated account of the transition from water-power to steam-power; Eliot's sensitivity to this interstitial moment between energy regimes becomes a methodological opportunity because it makes visible our need for a critical practice willing to toggle between past and present in order to grasp the scale of challenges facing us now. The concept of sustainability also originates as a problem of Victorian form, finds Deanna K. Kreisel, who points out that John Ruskin's writings on organicism productively fail to differentiate the living from the non-living. By defining life as ordered form, Ruskin's writings on seemingly inert natural objects—rust, crystals, and leaves—point the way to contemporary sustainability theory by immersing the human in the natural world and showing a dramatic dynamism to characterize both. Adam Grener rethinks the relationship between empire and ecology in Victorian realism, focusing on the crucial role of weather and atmospheric imagery in Victorian efforts to conceptualize systemic interconnection. If Charles Dickens's *Dombey and Son* (1846–48) fails at the level of content to correlate the Empire with its nationally scaled visions of reform, this novel's globalized ecological tropes nevertheless demonstrate how novel form cannot but situate the local and particular within the totalizing systems that contain them.

The book's third section, "Scale," shows how Victorian literary and theoretical writing engaged productively with the scalar distortions that followed from their efforts to comprehend vastly complex systems like ecologies and empires. These essays make the case for Victorian authors' self-conscious movement between registers of magnitude and their exploitation of what Bruno Latour calls the "zoom effect."[29] For Benjamin Morgan, utopian form can be defined precisely because of its scalar qualities: Committed to mediating totality, utopia is attuned to interactions between human and nonhuman systems at multiple levels. This capacity comes into focus when William Morris's *News from Nowhere* (1890) is read in light of Samuel Butler's *Erewhon* (1872), a satire of settler colonialism that recognizes the multiple levels—individual, societal, imperial—at which society and the economy is infused with nature and biology. Lynn Voskuil tracks the global ambitions of Victorian botanical science to show how scale emerged as a fundamental conceptual challenge for thinkers aiming to conceive life in systemic terms. For Voskuil, Joseph Dalton Hooker's struggle to account for the global distribution of plant species, and for the perspectival distortions of the landscape and biosphere he experienced in the Himalayas—recounted in his *Flora Indica* (1855) and *Himalayan Journals* (1854)—anticipate and foreground the scalar distortions inherent in more recent critical turns to "distant," quantitative methodologies. Scale effects also trouble the Victorian novel, Aaron Rosenberg points out: He shows how Thomas Hardy's invocations of romance and melodrama became formal strategies for evoking magnitudes of experience beyond the human scale of realism, geological time and astronomical space. In *Two on a Tower* (1882) and *A Pair of Blue Eyes* (1873), the scalar patterns native to realist form thwart the marriage plot, even as the sensational modes rising to fill their place prove capable of bringing deep time and space into alignment with the (human) present.

The book's final section, on "Futures," bears out our shared conviction that the century of coal remains our own. These essays show how this sense of continuity or even intimacy with our nineteenth-century past might productively distend the boundaries of period and nation still structuring humanistic inquiry. Resilience, persistence, and oppositional ongoingness: These and related figures for capacity structure the essays in this section, and demonstrate that the aim of this collection is not merely to write the disaster, but to think with it and through it. Not to rest in the often self-aggrandizing modes of elegy, witness, or sublime renunciation but to begin the work of imagining forms of life and work that might move us, together, toward livable futures. Monique Allewaert relocates Marx's famous commodity fetish—the degree zero for critical accounts of western modernity—and relocates this decisive concept in the Atlantic milieu whose African-American fetish practices gave it shape. In tracking the animist legacies of this critical concept, Allewaert shows how continents, cultures, and ecologies have intersected in the past to imagine forms of value not just within but outside exchange. Thus do the writings of Marx's contemporary Martin Delany, particularly *Blake; or the Huts of America* (1859–62), construe from those animist legacies an acapitalist mode of valuation that offers hope for the present moment: "It's time, again," Allewaert writes, "to be cheered by the strange movements on the edges of empire and the materialisms that flash forth from them." Hope also radiates from Teresa Shewry's account of satire's long arc across the history of settler colonialism, from Butler's *Erewhon* to the contemporary poetry of David Eggleton. If satire now seems to short-circuit in the face of ecological crisis, Shewry argues, its tone seemingly mismatched to the scale of its object, that may be because satire pinpoints our lingering affective attachments to fossil-fueled lifeways. In the context of such residual attachments, satire's capacity for scathing critique holds out the possibility of alternative futures, beyond those prescribed for us by habit—if we choose to take them. Karen Pinkus concludes the volume by putting our shared values of experimentalism, improvisation, and creative resilience into explicit practice. Her contribution takes the shape of a dialogue between Jules Verne's 1877 fantasy novel about coal extraction, *Les indes noires* (*The Black Indies*), and its twenty-first-century reader, "Karen Pinkus." This oscillation between nineteenth- and twenty-first-century subject and object generates a productive groundlessness, an interface between a fantasy tale from the coal-age and our own critical moment that for Pinkus yields some qualified push toward possibility. "We must take care of each other," she concludes. "[B]ut can we imagine doing so outside of . . . escapist fantasy?"

To finish, then, we rework that question by offering two canonical Victorian scenes that have come to haunt us as this project has taken shape. The first takes us back to *Tess*: It is that character's harrowing late-night baptism of her dying baby, the living reminder of her rape by Alec D'Urberville, whom Tess names Sorrow. Tess's hasty and theologically empty baptismal rite is meant to secure for her dead baby a future redemption that readers know it will not receive. We find in Tess's insistence on Sorrow's churchyard burial a startling refusal to abandon the project of care, despite the seeming futility of that commitment. Sorrow's life was so short, we are told, that he thought "the week's weather climate": All he knew was the weather of his own short time here.[30] In the midst of our own brief time on earth in the late Anthropocene, Tess flashes forth not just a disposition of persistence and fidelity amid catastrophe. In the face of the suffocating, attenuating systems that would render action null, Tess also refuses to abandon her conviction that individual works of care might, and do, matter.

The second scene is Robert Browning's "Love among the Ruins" (1855), which views the site where an imperial city once stood, but which has since been erased by nonhuman life:

> Now—the country does not even boast a tree,
>   As you see,
> To distinguish slopes of verdure, certain rills
>   From the hills
> Intersect and give a name to (else they run
>   Into one)
> Where the domed and daring palace shot its spires
>   Up like fires[31]

What catches our attention, and distinguishes this from so many other nineteenth-century visions of vanished empire and the forms that outlast them—from Shelley's "Ozymandias" (1818) to Dante Gabriel Rossetti's "The Burden of Nineveh" (1856)—is the haunting sense of a natural world gladly shrugging off its human traces. "O heart!," Browning writes. "[O]h blood that freezes, blood that burns! / Earth's returns / For whole centuries of folly, noise, and sin!"[32] Faced with this scene of desolation, the poem answers with a cliché, "Love is best."[33] Of all the tasks facing us now, one of them, we suggest, might be to strip the varnish from that maxim and from this poem, as from *Tess*, and to recover a more pressing and even radical form of love, expanded now beyond species division and even beyond the category of life. Doing so might help us imagine how, under an affect of care and solidarity, we might yet imagine possibility and co-evolution from amid the disaster of our present.

Notes

1. Freud notes, as though talking about the earth under capitalism, that "[t]he act of obtaining erotic mastery over an object coincides with that object's destruction." In healthy libidos, this originary sadism toward the object gets transformed for purposes of reproduction. Where this conversion toward generation fails, "we should produce an example of a death instinct," Freud says; but "this way of looking at things is very far from being easy to grasp and creates a positively mystical impression." Sigmund Freud, *Beyond the Pleasure Principle*, trans. and ed. James Strachey (New York: Norton, 1961), 65.

2. Jeremy Davies, *The Birth of the Anthropocene* (Berkeley: University of California Press, 2016), 5.

3. Ibid., 39.

4. Rob Nixon, *Slow Violence and the Environmentalism of the Poor* (Cambridge: Harvard University Press, 2011), 2.

5. Charles Darwin, *The Structure and Distribution of Coral Reefs, Being the First Part of the Geology of the Voyage of the Beagle, Under the Command of Capt. Fitzroy, R.N. during the years 1832 to 1836* (London: Smith, Elder, and Co., 1842).

6. Walter Benjamin, "Theses on the Philosophy of History," in *Illuminations*, trans. Harry Zohn (New York: Schocken Books, 1969), 249.

7. Amitav Ghosh, *The Great Derangement: Climate Change and the Unthinkable* (Chicago: University of Chicago Press, 2016), 62–63.

8. Dipesh Chakrabarty, "The Climate of History: Four Theses," *Critical Inquiry* 35, no. 2 (2009): 221.

9. Ghosh, *Great Derangement*, 63, 33.

10. As Karen Pinkus writes, "We might go to an extreme and suggest that whether or not we explicitly take up climate change in our writing (critical, creative, institutional-bureaucratic, or otherwise), climate change takes us up. Writing in the time of climate change—even critical writing engaged with texts from before the widespread extraction of fossil fuels—is necessarily untimely, out of joint with familiar modes of thinking and being, no matter how heterogeneous these may be." "Climate Change Criticism," *Diacritics* 41, no. 3 (2013): 3.

11. The challenges that climate change poses to humanistic method are summarized in Ian Baucom and Matthew Omelsky, eds., "Climate Change and the Production of Knowledge," Special issue of *SAQ: South Atlantic Quarterly* 116, no. 1 (2016); Ian Baucom, "History 4°: Postcolonial Method and Anthropocene Time," *Cambridge Journal of Postcolonial Literary Inquiry* 1, no. 1 (2014): 123–42; Chakrabarty, "Climate of History." Another important line of thinking, often adding feminist concerns to the sometimes masculine debates noted previously, follows Donna J. Haraway to show how the nature-culture division crucial to certain strains of humanities thinking breaks down under our climate crisis. See, for example, Haraway, *Staying with the Trouble: Making Kin in the Chthulucene* (Durham: Duke University Press, 2016); Anna Lowenhaupt Tsing, *The Mushroom at the End of the World: On the Possibility of Life in Capitalist Ruins* (Princeton: Princeton University Press, 2015); and Alexis Shotwell, *Against Purity: Living Ethically in Compromised Times* (Minneapolis: University of Minnesota Press, 2016).

12. Baucom and Omelsky, "Knowledge," 2.

13. The nineteenth century "saw the emergence of meteorology and climatology as two distinct fields," with the foundations of a "holistic concept of climate" laid by Alexander von Humboldt. Victorian scientists deduced the "fundamental energy-transport function of the climate system," while Svante Arrhenius first proposed in 1896 that carbon dioxide could produce atmospheric warming at global scale due to a greenhouse effect. Matthias Heymann, "The Evolution of Climate Ideas and Knowledge," *Wiley Interdisciplinary Reviews: Climate Change* 1 (2010): 587–89; Paul N. Edwards, "History of Climate Modeling," *Wiley Interdisciplinary Reviews: Climate Change* 2 (2011): 128.

14. Tobias Menely and Jesse Oak Taylor, Introduction to *Anthropocene Reading: Literary History in Geologic Times* (University Park: Penn State University Press, 2017), 6. We also note here the formation in 2016 of the "Vcologies" group, which formalized a gathering interest in Victorian ecocriticism and, in the manner of a Saussurean sign, did much to make a new object—Victorian ecology—conceptualizable.

15. For exceptions see Siobhan Carroll, *An Empire of Air and Water: Uncolonizable Space in the British Imagination, 1750–1850* (Philadelphia: University of Pennsylvania Press, 2015); John McNeill, *Mosquito Empires: Ecology and War in the Greater Caribbean, 1620–1914* (New York: Cambridge University Press, 2010); and Mike Davis's astonishing *Late Victorian Holocausts: El Niño Famines and the Making of the Third World* (London: Verso, 2002). That the latter book, in particular, has yet to be interpolated into Victorian Studies' self-understanding as a field is symptomatic of the divide between ecological and human histories of catastrophe we seek here to address.

16. Jason W. Moore, *Capitalism in the Web of Life: Ecology and the Accumulation of Capital* (London: Verso, 2015); Jennifer Wenzel, "Turning Over a New Leaf: Fanonian Humanism and Environmental Justice," in *The Routledge Companion to the Environmental Humanities*, ed. Jon Christensen, Ursula K. Heise, and Michelle Niemann (Abingdon: Routledge, 2017), 166.

17. Thomas Hardy, *Tess of the D'Urbervilles*, ed. Tim Dolin (London: Penguin, 1998), 187.

18. Ibid., 186.

19. Jason W. Moore, *Capitalism in the Web of Life*; Anna Lowenhaupt Tsing, *The Mushroom at the End of the World*. Jacques Derrida, "Structure, Sign, and Play in the Discourse of the Human Sciences," in *The Structuralist Controversy: The Languages of Criticism and the Sciences of Man*, ed. Richard Macksey and Eugenio Donato (Baltimore: Johns Hopkins Press, 2007).

20. Georg Lukács, *The Theory of the Novel: A Historico-Philosophical Essay on the Form of Great Epic Literature*, trans. Anna Bostock [1920] (Cambridge: MIT Press, 1974).

21. Raymond Williams, *The Country and the City* (Oxford: Oxford University Press, 1973), 18.

22. Ibid., 2. That nature and culture "coproduce" one another—indeed, that "we might instead look at the history of modernity as co-produced, *all the way through*"—is the insight of Moore, *Capitalism in the Web of Life*, 7, original emphasis.

23. For John Barrell, "Hardy attempts to communicate a notion of... [a] primitive [sense of] geography, as it inheres in its purest form in the consciousness that he attributes... to Tess before her marriage." "Geographies of Hardy's Wessex," in *The Regional Novel in Britain and Ireland, 1800–1990*, ed. K. D. M. Snell (Cambridge: Cambridge University Press, 1998), 112.

24. Hardy, *Tess of the D'Urbervilles*, 36, 37.

25. Ibid., 363.

26. "ecology, n." OED Online, Oxford University Press, January 2018, http://www.oed.com/view/Entry/59380?redirectedFrom=ecology (accessed March 16, 2018); Robert C. Stauffer, "Haeckel, Darwin, and Ecology," *The Quarterly Review of Biology* 32, no. 2 (June 1957): 138–44, 143. This account draws on Elizabeth Carolyn Miller, "Ecology" (Unpublished manuscript, March 3, 2018).

27. Miller, "Ecology," 2.

28. George Eliot, *Middlemarch*, ed. David Carroll (Oxford: Oxford World's Classics, 2008), 261.

29. Bruno Latour, *Reassembling the Social: An Introduction to Actor-Network Theory* (Oxford: Oxford University Press, 2005), 186.

30. Hardy, *Tess of the D'Urbervilles*, 96.

31. Robert Browning, "Love among the Ruins," in *The Norton Anthology of English Literature*, ed. M. H. Abrams and Stephen Greenblatt, vol. E, *The Victorian Age*, ed. Carol T. Christ and Catherine Robson, 8th ed., lines 13–20 (New York: Norton, 2006).

32. Ibid., lines 79–81.

33. Ibid., line 84.

PART I
Method

CHAPTER I

Drama, Ecology, and the Ground of Empire
The Play of Indigo
Sukanya Banerjee

In his thought-provoking exposition on climate change, *The Great Derangement: Climate Change and the Unthinkable* (2016), Amitav Ghosh, an anthropologist by training, leverages his formidable reputation as a novelist to dwell at length on the failure of our literary imagination. According to him, the realist novel, as it developed over the past two hundred years, expunged what our understanding of the Anthropocene has now alerted us to, namely a "renewed awareness of the elements of agency and consciousness that humans share with many *other beings* and even perhaps the planet itself."[1] For Ghosh, the realist novel, in its preoccupation with the routinized everyday of bourgeois life, progressively unlearns this intimacy of shared agency and consciousness, casting it to the realm of the "unheard-of" or the "unlikely."[2] Casting a wide net, he discusses Gustave Flaubert's *Madame Bovary* (1857) and Bankimchandra Chatterjee's *Rajmohan's Wife* (1862) as examples, suggestively adding that, "It is as though our earth had become a literary critic and were laughing at Flaubert, Chatterjee, and their like, mocking *their* mockery of the 'prodigious happenings' that occur so often in romances and epic poems."[3]

There is much that is persuasive about Ghosh's claims, and one cannot, of course, turn away from the urgency of his overall argument. But his literary assumptions also invite debate. One can, for instance, take issue with Ghosh's conflation of the literary imagination with a novelistic one, or with the relation that he etches between realism and genre fiction, or, indeed, with his reading of *Rajmohan's Wife* itself inasmuch as that novel is shot through with elements of the gothic and supernatural.[4] Set in nineteenth-century rural Bengal, *Rajmohan's Wife*—the first Indian novel in English—deals with the imputed infidelity of the eponymous character (it is not because of its realist aspirations alone that Ghosh pairs it with Flaubert's classic). I draw attention to *Rajmohan's Wife* because this essay examines the literary landscape that Chatterjee's novel inhabits. By reading Dinabandhu Mitra's play, *Neel Darpan* (1860), which was contemporaneous with Chatterjee's novel, this essay discloses a more mottled—and contested—site of literary production than Ghosh amply intimates. It is worth noting that *Rajmohan's Wife* was barely read when it was published in installments in the journal *Indian Field*.[5] On the other hand, *Neel Darpan*, which is a play about indigo cultivation, captured the popular imagination soon after it was published and attracted an enthusiastic audience wherever it was performed. Acknowledging this varied literary milieu, I suggest, is significant for our own methodologies, not least because it brings drama (particularly that of colonial provenance) into the ecocritical conversation in ways that expand the formal as well as geoimperial scope of that conversation.

Interestingly, Ghosh has very little to say about drama, which was a highly popular form in mid–nineteenth-century Bengal (though, to be fair, he admits that as a novelist, he is drawn to discussing the form closest to his heart). Of course, Ghosh is not the only one guilty of overlooking drama. Drama, by and large, has received short shrift in Victorianist scholarship as well. Redirecting attention to drama (to include both play and performance) does not only recompense for the lapses in our scholarship but, as this essay argues, also shines light on the dramatic form as one that keeps alive the sense of "shared agency and consciousness," whose loss in an individualized, novelized modernity Ghosh laments. It is telling that in his formulation of actor-network theory (which Ghosh cites for its salutary undoing of the Cartesian divide), Bruno Latour seems to rehearse the classical principles of drama. To be clear, Latour does not use "actor" in the conventionally used dramatic sense of the term; for Latour, "actor" extends well beyond humans to include any entity—human, nonhuman, unhuman—that "acts or to which activity is granted by others."[6] But if for Latour, the "actor"—or actant—qualifies for that designation by being the source of action or "doing,"[7] then it is worth keeping in mind that as Aristotle notes in the *Poetics*, "drama" originates from the word *dran* in the Megarian dialect, which means "doing."[8] "[In drama]," Aristotle points out, "agents accomplish the imitation by acting."[9] Here "acting" seems to connote representation as much as it connotes action, and Aristotle accords a higher level of action to drama than to epic poetry.[10] Significantly, when rethinking human-nonhuman networks on the basis of an entity's "acting," Latour often resorts to vocabulary that is redolently of and from drama.

One objective of reading *Neel Darpan*, then, is to underscore the salience of drama to the ecocritical imperative of drawing attention to a "multiply centered expanse" in which humans are not the only agentive entities.[11] After all, as Baz Kershaw notes, the very process of staging drama underlines the extent to which drama is constituted by "unavoidable interdependencies between every element of a performance event and its environment," which makes "theater ecology a matter of living exchange between organisms and environments."[12] But while this point could perhaps be made through the analysis of just about any play, to read a mid–nineteenth-century Bengali play about indigo cultivation is also to bring home the materiality of empire to the study of Victorian ecology. *Neel Darpan* details the brutal effects of forced indigo cultivation in lower Bengal and Bihar. Indigo was indigenous to the Indian sub-continent and was grown and processed mostly in the western part of the country. In fact, the first commercial venture of the East India Company (EIC) in Surat in the seventeenth century consisted of a highly profitable investment in indigo.[13] Over the seventeenth century, however, indigo cultivation moved to the West Indies, where European planters began producing a superior quality of indigo that fulfilled the high demand for indigo dye in the European market. But the planters in the West Indies soon diverted their attention to growing even more profitable crops such as sugar and coffee, and the cultivation of indigo moved by the mid–eighteenth century to southern Carolina, Spanish Guatemala, and French Santo Domingo. The outbreak of the American Revolutionary War, in turn, made it difficult for the British to access their trade routes for indigo, and the East India Company began to revive its interests in the crop. But this time it decided to grow indigo in Bengal and Bihar, where the company had by then established a stronghold but where indigo had never been cultivated before. Nonetheless, by 1842, indigo accounted for as much as "forty six per cent of the value of goods exported from Calcutta."[14]

If this brief snapshot of the trade and cultivation of indigo tells us anything, it is that the peripatetic fortunes of colonialism and the indigo plant are inextricably intertwined, and it is impossible to bifurcate social history (of colonialism) from environmental—or even botanical—history (of indigo and its cultivation). Simply put, colonial and environmental histories are interdependent, a fact that, as scholars such as Ramachandra Guha and David Arnold, Elizabeth DeLoughrey and George Handley, and Rob Nixon, among others point out—and as Nathan Hensley and Philip Steer note in the Introduction to this volume—mainstream Anglo-American ecocriticism has often failed to take due note of.[15] The emphasis on "history in nature" rather than on a dichotomous view of "history and nature," is also, as we know, crucial to Jason Moore's formulation of the "ecological" as offering "a holistic perspective on the society-environment relation." But for Moore, such a perspective also questions how "master processes of colonialism, etc." remain "resolutely social," always "ceded to the Cartesian binary."[16] What this ecological reformulation makes obvious for scholars of Victorian studies is that if over the last two decades we have reached a stage in which it is difficult to absent the history of empire from that of Victorian Britain, then we are now also at the stage where we cannot speak of "empire" in terms of its human constituency alone.[17]

Therefore, if the "imperial turn" prompted an interest in the contact and engagement between Britons and colonial peoples, then an "ecological turn" calls for an understanding of the multiple relationalities not just between colonizer and colonized, but between the human and nonhuman, "society and environment," in Britain and beyond. In this, though, the idiom of mobility, which was key in conceptualizing the imperial turn, is not the only one that is key to expanding our sense of empire. Rather, as this essay argues, an idiom of "groundedness" becomes equally crucial, for we may very well be speaking of entities, objects, and collectivities that literally do not move. Drama, as perhaps the least readily mobile of literary forms (if we take its individual performances into consideration, that is), serves as both a heuristic and an exemplar for a critical methodology that can give due accord to a logic of groundedness.

In reading *Neel Darpan*, therefore, this essay is attentive to its dramatic features and also takes the play as well as its performance history into consideration. Such an emphasis enables the essay to shift the focus to indigo, which is an integral component of the play but barely receives critical attention, given that much of the scholarship on the play is interested in its purported anticolonialism (the play depicts the popular resentment against the planters, which fueled the "indigo rebellion," a series of protests that broke out in different areas of lower Bengal in the late 1850s). Such interest is not misdirected; *Neel Darpan* certainly spoke to an incipient nationalism and was chosen for the inaugural performance of the National Theatre in Calcutta in 1872.[18] But if the objective of this essay is to emphasize the importance of drama to ecology as well as to twin ecology with colonialism, then it becomes something of an imperative to be attentive to the long-neglected role that indigo itself takes on in the play (in ways that are not unrelated to the play's politics). Such a reading addresses the urgency underpinning the questions that Michael Taussig asks in a broader context about our seeming obliviousness to the vibrancy of indigo as an entity:

> Is it not time for blue [indigo] to exert its magic and sexuality ... so as to undo that which would cast it as "color," sans history, sans density, sans song? If it could penetrate an egg and make men cough blue, this beauty that is indigo, how much more likely is it to penetrate history as a silent symbol ensconced in a color chart? When will we cough blue?[19]

In what follows, I offer an account of indigo cultivation in Bengal and the events leading up to the "indigo rebellion" that prompted the play; then, after placing *Neel Darpan* in its colonial setting, I show how its overlooked dramatic features foreground the role of indigo in ways that not only enhance the reading of the play but also emphasize the ecohistorical nature of the event that the play charts; finally, I consider how the "groundedness" of drama might be important for twinning ecology with empire in ways that may well make us, in Taussig's terms, cough blue.

The Play

When the East India Company revived the indigo trade in India in the late eighteenth century, it invited European planters, many of whom had owned and managed plantations in the Caribbean, to take up indigo plantation in Bengal. Much of the indigo was cultivated in villages by peasants (*ryots*) under contract with the planters who paid them an advance, thereby obliging them to produce a certain amount of indigo. The land on which the *ryots* grew indigo was land over which they had tenancy rights; this was either land that the planters had leased from local landowners or land that was owned and managed directly by the landowners. Cultivating indigo, however, was simply not profitable for the *ryots*, and they barely earned enough to recover costs.[20] That being the case, it was very common for *ryots* to be coerced into signing contracts or to consent to contracts they did not fully understand. Planters maintained their hold over the *ryots* through a combination of physical violence and intimidation and went so far as to kidnap and detain them if they did not produce enough indigo.²¹ Local landowners were also forced to plant indigo on their lands, with the result that many of them supported the *ryots* when they protested the planters' actions.

To be sure, the East India Company took a dim view of the coercive labor practices deployed by the planters, but it was equally convinced that professional planters alone could generate the maximum profit. Therefore, over the first few decades of the nineteenth century, the Company was hesitant to take the planters to task despite regular complaints about their tactics. From the mid-1850s onward, however, there were reports of peasant unrest in the indigo-growing districts of lower Bengal. Evidently, even as the famed revolt of 1857 spread across north India, commandeering political and media attention due to the spectacular nature of its events, unrest about other matters was quietly fanning micro-rebellions in other parts of the country as well.

Matters came to a head in 1859–60, and the following account (albeit a retrospective one) captures the general mood of the time:

> Europeans riding about the country were insulted and assaulted. Planters were violently resisted in the performance of their usual works, such as measuring lands; . . . Growing crops were destroyed. Factories began to be attacked and plundered, and in some cases burnt. . . . Mobs assembled in large numbers, armed with spears swords bamboos and shields [*sic*].²²

Taking due note of the accelerating pace of events, Lord Canning, Viceroy of India at the time, reportedly commented, "I assure you that for about a week, it caused me more anxiety than I have had since the days of Delhi."²³ The government appointed a commission to prepare a report on the causes for the "indigo rebellion" and on the status of the indigo industry in Bengal. The report, submitted in August 1860, was critical of many of the practices followed by the planters even as it recommended steps to safeguard them against heavy losses. Nonetheless, official denunciation of their exploitative practices did pave the way for more effective judicial access and protection for the *ryots* in ways that curtailed the planters' oppressive tactics, and eventually, indigo cultivation itself.²⁴

Significantly, the commission's report was submitted only a few weeks before *Neel Darpan* was published. It was a common perception of the time that the play did as much as, if not more than, the report to draw attention to the indigo problem. Dinabandhu Mitra, the playwright, was a government official. As a Superintendent under the Post Master General, Bengal, he had toured and lived in the indigo-growing districts and was well acquainted with the uncongenial practices attending indigo cultivation. Although he went on to write plays that commented on contemporary social topics by combining romantic comedy with satire, the unrelenting bleakness of *Neel Darpan*, his first play, seems most informed by the immediacy of his experiences in the indigo-growing areas. *Neel Darpan* is pointedly critical of the planters. This stance in itself is not unique, for urban Bengali-language theatre was in some measure built around a politics of social protest. Ramnarayan Tarakratna's *Kulinkulsharbashwa* (1857), reputedly the first Bengali play written for the stage (in contrast to Bengali plays that were adaptations of Sanskrit or English plays), was in fact commissioned to protest the polygamous practices of upper-caste Brahmins, and the genre of "protest plays" continued well into the 1870s.[25] But *Neel Darpan* may have been the first work of its kind to directly implicate Europeans, a fact that also seems to have bestowed on it instant notoriety. James Long, an Irish missionary belonging to the Church Missionary Society, had the play translated into English (reportedly by noted Bengali poet Michael Madhusudan Dutt) and then distributed copies amongst high-ranking officials in both India and England by way of acquainting them with the plight of the *ryots*.[26] The translated edition was published as *Nil Darpan: The Indigo Planting Mirror*. Predictably, the circulation of the translated copies met with the planters' disapproval, and they sued Long for libel. Long was sentenced to a month's imprisonment and fined a thousand rupees, and the circulation of translated copies of the play was severely restricted.

Although the charges were brought against Long and not Mitra, the playwright, it is not hard to see why the translation provoked such pushback from the planters, or why its original version met with such popular enthusiasm. Set in the village of Swarpur, the play centers on the fairly prosperous family of Golak Chandra Basu, the landowner. Golak Chandra is resentful of the planters' demand that he grow more indigo on his land; the planters in turn file a false case against him, accusing him of inciting the *ryots* to rebel. The planters collude with the magistrate to have Golak Chandra jailed. Unable to bear the shame of imprisonment, Golak Chandra commits suicide, which in turn triggers a series of events bringing about the death of most of his family members, including his older son, Nabinmadhab. Golak Chandra and Nabinmadhab are cast as benevolent, if paternalist, figures, quite unlike the two English planters, P. P. Rogue and I. I. Wood, whose sole interest in extracting maximum profit from the land is marked by their invective-filled speech and habitual acts of physical violence directed toward the *ryots* and Golak Chandra's family.[27]

The planters' violence takes up considerable stage-time. Almost every utterance that Wood and Rogue make is laden with expletives or threats. Within a few moments of his appearance in the play, Wood boasts of his "money, horses, musclemen, spearmen" with which he intimidates the peasants into sowing indigo on their land.[28] As he tells his overseer: "I whipped those mother-fuckers, snatched cattle, locked up the wives" (Mitra 1.3, p. 191). The dialogue between the characters in several scenes of the play provides graphic descriptions of the many tactics used by the planters. The following exchange takes place between peasants who had been imprisoned for refusing to give false testimony against Golak Chandra:

> FIRST PEASANT: They won't keep us in one piece if we don't give evidence; Wood *saheb* stamped on my chest—see the blood's still streaming down. The swine's feet are like the hooves of a plough-ox.
>
> SECOND PEASANT: Those were nails, sharp nails—didn't you know *saheb*s wear nailed boots? (2.1, p. 203)

Even female members of the peasants' families are not spared. In a particularly melodramatic scene, Rogue attempts to rape Kshetramoni, the pregnant wife of one of the peasants, and when she (successfully) resists him, he swears: "Shut up you bitch, mouthing such big words!" (2.3, p. 226).[29] The accompanying stage-directions for these lines indicate "[*He pulls her by her hair and punches her in the abdomen*]." These directions must have been faithfully executed, for at the play's staging at the National Theatre, the prominent social reformer Ishwarchandra Vidyasagar allegedly flung a shoe at the actor playing Rogue.[30] At the play's staging in Lucknow in 1875, Rogue's exaggerated depiction annoyed a few Englishmen in the audience who then tried to beat up the actor standing up to Rogue's villainy (this would be Nabinmadhab, who enters the scene at this point along with Torap, a *ryot*).[31]

In describing the "melodramatic imagination of nineteenth-century nationalist playwrights," Nandi Bhatia notes how depicting the rape or torture of Indian women by white men "served as a powerful mode of representation for evoking the audiences' emotions and displaying the magnitude of colonial oppression."[32] In this, *Neel Darpan* is no exception except that the emotional response that it elicits against the planters is as much a result of their fundamental reordering of the *ryot*/landowners' relation to the land as their physical violence. In fact, while the planters' acts of violence literally occupy center stage (as in the scene of attempted rape), the less spectacular but equally corrosive violence of the planters claiming indigo land is what the play's plot hinges on.[33] A disturbing consequence of this is that not much is made of Kshetramoni's attempted rape after the incident, and her fate in the play (she dies) is readily linked with that of other characters, who also meet a similar end because of the planters' dismantling of their livelihood.[34] What this also means is that the land does not necessarily become metaphoric of the woman's body (or vice versa)—a favored nationalist tropology—but is depicted more in its materiality. In fact, the play opens with Golak Chandra complaining about how the planters have made incursions into the land near his house: "They've ploughed up the land all around the pond, they will sow indigo there" (1.1, p. 186). The planters' practice of sending emissaries to forcibly stake the land for indigo cultivation upon payment of a cash advance is what is most despised and is referred to throughout the play.

Significantly, it is indigo, associated with the excesses of a colonial-planter economy that alienates the peasant-cultivators from the land, that becomes metaphoric of a destructive colonial presence. At a later point in the play, when Nabinmadhab reluctantly contemplates pawning his wife's jewelry so that he can pay for the legal costs for defending his father from the false charges of the planters, he vents his deep-seated frustration at the planters by railing against the condition that indigo cultivation has reduced him to: "Ours was such a happy family—what was I, and what am I today! From our property our profit used to be seven-hundred rupees a year, fifteen barns bursting with rice. The orchard alone covered fifteen bighas..." (3.2, p. 220). Indigo even becomes synonymous with the planters, who are commonly referred to as "indigo-devils." As one of the tenants points out to Golak Chandra's wife, "Ma, though half-starved, we're sowing *their* indigo. All the bighas *they* mark out, we're sowing on them all" (3.2, p. 223; italics mine).

The play's overt identification of the planters with indigo and vice versa is noteworthy because it underlines the inextricability of colonial and environmental histories.[35] While I will develop that point further in terms of its ramifications with reference to studies of empire, for now I want to consider Elizabeth DeLoughrey and George Handley's observation that "[s]ince it is in the nature, so to speak, of colonial powers to suppress the history of their own violence, the land and even the ocean become all the more crucial as recuperative sites of postcolonial historiography."[36] For DeLoughrey and Handley, "one vital aspect of postcolonial ecology is to reimagine [the] displacement between people and place through poetics."[37] Read through such a lens, *Neel Darpan*, despite its tragic conclusion, can be seen as offering a final note of redemption. In the concluding scene, one in which almost every member of Golak Chandra's family dies, his younger (surviving) son, Bindumadhab, is given the closing lines. In mourning his family through perhaps the most stylized language in the play, Bindumadhab invokes the land in terms of its natural beauty, returning it to a state of plenitude that seems to recompense him for his loss and imbue him with a sense of divinity:

> Man comes into this world for a while to play his part—as ephemeral as the high banks of a swift river. Yet how beautiful is the bank of the river—covered with soft green *darba* grass, with mighty trees bedecked with fresh leaves . . . A strole [*sic*] there, the sweet songs of the birds ringing in one's ears, the flower-scented breeze fanning one—all these fill one with the thought of that Joyous Eternal Being. (5.4, p. 266–27)

In their calm cadence, these lines offer a sharp contrast not just to the planters' fantastic violence, but to their staccato speech as well. They also segue into the final section, which, as the only extended section of the play to be set in rhyme (in the Bengali version), seems marked off from the play and functions as an epilogue.[38]

The epilogue comprises a series of short verses through which Bindumadhab narrates the main events of the play. The opening verse of the epilogue states:

> The Indigo Planter is a venomous snake,
> My happiness he has reduced to ashes.
> Injustice killed father in the prison,
> Brother was felled in the indigo field.
> Grievous loss made mother mad, and
> In a fit of madness lovely Saralata she killed. (267)

The musicality of the "epilogue" (evident in the Bengali version), as well as its narration of a sequence of events through verse, recalls the popular folk forms of *jatra*, in which musical compositions played a prominent role. While the folk theatrical form of *jatra* had its origins in village festivals, by the middle of the nineteenth century it had seeped into the metropolitan culture of Calcutta as a result of mass migration to the colonial city.[39] Interestingly, urban Bengali theatre, as it was being shaped at the time, tried to distance itself from *jatra*, which was marked as both rural and lower class. Cannily mindful, however, of the heterogeneous urban population, the actor-producer Girish Ghosh, doyen of the nineteenth-century Bengali stage, defied bourgeois literary tastes by including musical compositions reminiscent of *jatra* into play scripts as a means of appealing to a broader audience.[40] In evoking the musicality of *jatra* (but not replicating it), the final section in *Neel Darpan* reconnects Bindumadhab to the land; in so doing, it evidently connects the play and its performance with a wider audience as well.

Indeed, it is difficult to miss the connection that the play had evidently formed with a growing theater audience given that, twelve years after its original publication and ten years after indigo cultivation itself had ceased and protests against it were long over, it was this play that was chosen to be staged at the opening of the National Public Theatre in Calcutta, which was the first public Bengali theater company (earlier theatrical productions were privately sponsored). This fact, coupled with the play's purportedly anticolonial stance, does make it difficult to desist from reading the play *and* its performance as offering a counterpoint that recuperatively reroutes the relation between the land and its colonized inhabitants. Yet, reading in such recuperation is perhaps too easy; such recuperation is also perhaps too pat.

For one thing, it glosses over the very different relation that the landless *ryots* share with the land as compared to the one forged by the landed Basus (after all, it is Bindumadhab who is reconnected to the land; we know little about what happens to the *ryots*). Also, the recuperation can only be read as such by following a line of argument that arrives at closure by reconnecting the "native" to the land—a reading that Bindumadhab's closing lines may very well proffer. Such closure seems a welcome one, especially in a colonial setting that is anyway beset by physical brutality (such as that of Rogue and Wood). Reading in such closure would be one way to respond to DeLoughrey and Handley's call to "reimagine [the] displacement between people and place through poetics." But that such a move would be inadequate at least in terms of the play is evident in the fact that *adivasis*— aboriginal groups inhabiting the Indian subcontinent—remain shadowy figures mentioned only in passing, either as figures to be exploited (like the land) or as figures that meld into the landscape. The Bengali and English versions of the play (from 1860 and 1862, respectively) offhandedly refer to the tribal figures as "*Buno*," which, in Bengali, means "of the jungle" and is also the name of the tribe inhabiting that region (many *adivasi* tribes occupied forested areas, hence the name, perhaps). Folded in with the land/jungle, the *adivasi* figures have no bearing on the events of the play; they seem to have no problem either with the loss of land or the brutalities of indigo cultivation (although, historically, during the manufacturing season, itinerant *Buno* laborers were recruited by contractors to process the indigo in the factories).[41] In other words, if Bindumadhab's reconnection with the land at the end of the play seems to call for a recuperative reading, the environmentalism undergirding it would be naively "green," or what Timothy Morton describes as blindingly "bright green."[42] Turning our attention to indigo, on the other hand, proffers a different picture, a sharper (if not differently colored!) reading of the otherwise undifferentiated "land," "nature," or "native," that a poetics of rehabilitation *can* lapse into even in a postcolonialist vein.

Indigo

Indigo, simply put, is everywhere in the play. Its unsettling effect is evident in its appropriation of the familiar sites of village-life: from the jail, which is now referred to as the "indigo-lock up," to the school children's song, "Oh dear candy seller, candy seller have you / steeped the indigo, the indigo" (2.3 p. 212). Significantly, there is a double valence to how indigo operates in the play. On the one hand, indigo, as mentioned in the previous section, is clearly associated with the English planters and even functions as a metaphor for colonial presence. But indigo also comes into its "own" in the play, quite independent of the planters. Rogue states at one point: "But we're not really bad people as such, it's our work with indigo which brings out the evil in us" (3.3, p. 224). The peasants dislike planting indigo not only because of the planters' brutish tactics but also because of the incessant care that indigo as a plant requires of them. One of the tenant-farmers, Sadhu Charan, who has twenty *bighas* of land at his disposal, describes how "Indigo needs four times the care paddy does, so if I have to grow indigo on the nine *bighas* marked, all my eleven *bighas* will lie fallow" (1.3, p. 193). Indigo indeed demanded a high degree of attentiveness. For one thing, it required impeccable timing: It had to be sown before the spring rains, harvested right after the autumn rains, and transported to the factory almost immediately after harvest lest it begin to ferment.[43] It also required incessant weeding. And because the best indigo was produced by seeds that were sown in April—also the time for sowing rice (the staple food crop in lower Bengal)—growing indigo meant that the cultivators would have to do so at the cost of imperiling one of their basic needs: food. Understandably, many of them were reluctant to convert their "rice land" into "indigo land."[44] As Sadhu Charan's brother, Ray Charan, asks in the play: "If indigo *eats* up all the five *bighas* in Sampaltola, how the hell am I going to feed my wife and kids?" (1.2, p.188; italics mine). Evidently, indigo took over (and away) the lives of its cultivators. Toward the end of the play, one of the characters greets the news of the landowner Golak Chandra's death with the comment: "The cursed indigo has eaten the old man and it won't be long before it eats up the old woman also . . ." (5.1, p. 243).

Not featured as an actor in the play but certainly functioning as an actant, indigo is endowed with agency and is the substance that drives the play.[45] Indigo determines the play's dramatic arc, ensuring a tragic outcome for its characters. To be sure, indigo is not listed amongst the dramatis personae of the play, but precisely because it is not—and here dramatic form becomes apropos—its effects are thrown into sharper relief given the "concentration of effect" that drama produces.[46] On an otherwise sparse stage, it is indigo that is the subject of extended dialogue between the characters; it is indigo that is the addressee of their soliloquies. In other words, it is difficult to "take our eyes off" indigo even if it is not on stage because it is so clearly the cause of what does take place on stage (and unlike the realist novel, there is no anthropocentric narrator to deflect our attention away from indigo, either).[47]

Una Chaudhuri notes how objects have long assumed importance on the realist stage: Nora's Christmas tree, Laura's glass menagerie, or Willey's suitcase, according to her, become "characters" that "exercise a direct, unmetaphorical power in the formulation of the dramatic action."[48] While one agrees with Chaudhuri's observation of theater's ability to dramatize objects, it also does not seem enough, or even appropriate, to say that indigo operates as a character in *Neel Darpan*, for such a designation would deliver indigo from an object status; it would also overlook indigo's absent-presence in the play. It seems important to avoid the former and hold on to the latter if only because doing so helps think of indigo more as an actant, as a bundle of actions "long before [it is] '*characterized*'; at which point competences begin to precede and no longer to follow performances."[49]

By bringing to the fore indigo's agentive (rather than anthropomorphic) aspect, then, *Neel Darpan* pivots on indigo's intimacy with the characters and on its imbrication with their biorhythms in ways that signals to us the otherwise bleached landscape that the overanimation of the human yields. "[T]he real mystery in relation to the agency of nonhumans," Ghosh reflects in *The Great Derangement*, "lies not in the renewed recognition of it, but rather in how this awareness came to be suppressed in the first place."[50] It is hard to fathom what Mitra, or his audience, thought of indigo. But the heightened audience engagement with certain "actors," such as the planters, when contrasted with the conceptual no-place accorded to, for instance, the tribal figures in the play, gives one indication of how the play and its performance overtly exacerbates the human-nonhuman divide, an emphasis on familiar dualisms that can be attributed, amongst other things, to a colonial fashioning of rationalized knowledge-systems.[51] It is equally significant, though, that the play also incorporates folkloric elements of *jatra* that bear an angular relation to such knowledge-systems inasmuch as *jatra* and other popular modes of performance hearkened to traditions of storytelling that regularly dallied with gods and goddesses, traversing the divide between the secular and the divine, the quotidian and the fantastic.[52]

But the challenge for us as readers is not to invest folkloric or precolonial forms with a salvific force that retrieves nonhuman agency. For one thing, such a romanticizing move would repeat the colonial one of positioning the global south on one side of the Cartesian divide (even as we seek to undo it).[53] Besides, "retrieval" or "rescue" themselves come as highly loaded colonial tropes. The challenge lies, then, in acknowledging the "messiness" of *Neel Darpan*'s terrain, such that a reading of the role that indigo plays is not bracketed off as extraneous to the play's anticolonialism—depicted through its caricatured, hyper-agentive planters—not least because colonialism, far from being a purely social process that only impacts "Nature" or "the environment," is itself, as Jason Moore insists, irreducibly "socio-ecological": It traffics in and produces multiple relations between human and nonhuman nature.[54]

But if colonialism is not just about the social, and there is no "Nature" that can easily be summoned in the name of the ecological, then how can we, as readers of the nineteenth-century British empire, track the socioecological? At one point in *Neel Darpan*, a peasant who is in the "indigo-lock up" cries out: "Oh indigo, indigo, did you come to this land to destroy us!" (2.1, p. 206). On the one hand, indigo could just stand in here as proxy for the planters, or as a metaphor, as mentioned earlier, for an alien colonial presence. On the other hand, the highly personal mode of address marking the appeal also helps pry apart the multiple relations that indigo forges between itself, the land ("rice" or "indigo"?), the planters, the landholder, his family, the peasant cultivators, and their families—relations that otherwise congeal (and disappear) in the name of the colonial. That these relations are primarily negative ones (between the planters and the peasants) makes it easier for them to be subsumed by the insalubriousness generally conveyed by the term "colonial," whose effects are often mistakenly read—and dismissed—as cause. In other words, one can very easily say that it is colonialism that causes the peasants' misery. To be sure it does, but rather than have that point preclude further analysis (or impel one that proceeds along a recuperative trajectory), it is perhaps more productive to note how colonialism functions as such only because it enters into and brings about a certain relation among the peasants, the land, the planters, and indigo. And it is precisely a careful disaggregation of these multiple and looping threads of relationality, I argue, that is critical to understanding colonialism in its socio-ecological cast. Ironically, A. Sconce, a judge in the indigo-growing district of Nadia, had written to Lord Beadon, Secretary to the Government of Bengal, cautioning: It is "no longer enough to measure the advantages of European capital and energy by the value of our exports of Indigo." Pointing out that a "Native landholder would shrink from the approach of indigo cultivation as they do from fire in the dry prairies of America," he noted how "the strong sentiments and warm feelings of the people are not being sufficiently investigated and discussed."[55] The letter was written (and ignored) as early as 1854. But the opportunity for the kind of socio-ecological analysis that Sconce's letter clearly advocates is lost, however, as soon as indigo becomes a commodity.

As mentioned earlier, the indigo harvest, as soon as it was reaped, was transported to the factory (owned by the planters) to be manufactured into dye. Comprising three stages, the manufacturing process involved steeping the leaves and branches in water in a vat and then transferring the yellowish-green runoff to a "beating vat," where the liquid was oxidized, and a precipitate, "indigo," settled in the bottom of the vat. The precipitate was then scooped out, cleaned, and boiled. After being washed and dried, the indigo was shaped into square cakes, which were stamped with the initials of the manufacturer and the date of production and then transported to auction houses in Calcutta. From Calcutta, the bars of indigo embarked on a spectacular journey as they were transported to England, and then, in turn, across Europe and the Empire.[56]

As a commodity—a bar stamped with the initials of the manufacturer and the single date of production—indigo is obviously abstracted from its conditions of manufacture: the steeping, beating, washing, and drying. In fact Taussig, enraptured by the animacy of indigo, focuses on the "beating" vat, delighting in the interaction of the Buno laborers with the swirling pool of indigo. To be sure, the commodity form elides the labor of the migrant Buno workers, who spent hours stirring the indigo in the "beating vat." But, as Margaret Ronda and Tobias Menely remind us, the commodity form also elides the "material prehistory" of the "ecological substance."[57] In other words, it is not only the Buno laborers who get abstracted, but also the relational bearing of indigo to its human-nonhuman environment. It is significant to point out here that indigo cultivation and manufacture were distinct processes, but even Taussig, who wants us to "cough blue," focuses only on the latter. Evidently, the commodity and its making displaces the plant and its ecologies.

Notably, it is *Neel Darpan* as well as its performance that alerts us to indigo's material prehistory not only by detailing the plant's fraught relation with the cultivators but also by reminding us, quite literally, of the scene of its cultivation. Significantly, after the play was performed in 1872, a review in the Calcutta daily, *Amrita Bazar Patrika*, stated that it "should be staged outside Calcutta—like Krishnanagar, Berhampore, Jessore where the contents would be more relevant and valuable"[58] (these places were located in the former indigo-growing districts). The compensatory gesture of the play lies in its narration of indigo cultivation: Narrative, after all, has long been viewed as a counterpoint to the abstractions entailed by the commodity form.[59] But, more important for our purposes, the play's compensatory gesture lies in the relationality that it *continually* invokes and establishes between its audience, indigo, and the land. "Oh indigo, indigo, did you come to this land to destroy us!" (2.1 206).

Grounded

Scholars writing on drama have long noted how drama itself is an ecological form par excellence.[60] In the context of the present argument, I am interested in the possibilities that drama opens up to consider "groundedness" as a critical idiom and practice. This is not so much about drama evoking a sense of "place" as about the fact that dramatic enactment literally "grounds" us in the site of performance, even if we are only reading a play script.[61] This does not mean, of course, that drama does not travel, or that mobility is in itself a bad thing. But this is to rethink how "groundedness" offers a vantage point crucial for limning the connections and reconnections between human and nonhuman, nature and history (or, nature in history). Not coincidentally, Dana Luciano and Mel Y. Chen begin their absorbing analysis of the imbrication of human and nonhuman bodies by studying Laura Aguilar's photograph of a woman imperceptibly blending in with an enormous boulder: The photograph is titled "Grounded #114."[62] A call for "groundedness" of course necessitates a heightened vigilance against reductive claims of nativism or autochthony, especially because in contrast to an earlier critical mood that deterritorialized "place" in the name of "space," we now find ourselves returning to "place" in its materiality—to a "regrounding," as Latour has recently put it—in the name of anthropogenic exigency.[63] The "groundedness" of drama in the context of the globalizing aspirations of the nineteenth-century Empire may well help reorganize our critical impulses, ensuring that we guard against nativism but also reconstellate what, literally, constitutes the "ground." Such a double move can, at the very least, remind us that indigo is not only a commodity that can make us cough blue but also a very busy plant with demands of its own.

Notes

An abridged version of this chapter originally appeared in *Victorian Studies* 58, no. 2 (Winter 2016), and is reproduced with permission from Indiana University Press.

1. Amitav Ghosh, *The Great Derangement: Climate Change and the Unthinkable* (Chicago and London: University of Chicago Press, 2016), 63, emphasis added.
2. Ibid., 16.
3. Ibid., 26, original emphasis.
4. For a reading of *Rajmohan's Wife* as a sensation novel that incorporates these elements, see Sukanya Banerjee, "Troubling Conjugal Loyalties: The First Indian Novel in English and the Transimperial Frame of Sensation," *Victorian Literature and Culture* 42, no. 3 (September 2014): 475–89.
5. S. K. Das, *The Artist in Chains: The Life of Bankimchandra Chatterji* (New Delhi: New Statesman Publishing Company, 1984), 21.
6. Bruno Latour, "On Actor-Network Theory: A Few Clarifications Plus More Than a Few Complications," *Soziale Welt* 47, no. 4 (1996): 369–81, 373. For Latour's detailing of actor-network theory, see his *Reassembling the Social: An Introduction to Actor-Network-Theory* (Oxford: Oxford University Press, 2005).
7. See, for instance, Bruno Latour, "Agency at the Time of the Anthropocene," *New Literary History* 45, no. 1 (Winter 2014): 1–18.
8. Aristotle, "Poetics," in *Classical Literary Criticism: Translations and Interpretations*, edited by Alex Preminger, et al (New York: Frederick Ungar Publishing, 1974): 97–139, 110.
9. Ibid., 113.
10. Ibid., 114. For a more extensive consideration of Aristotle's association of drama with action, see Gerard Genette, *The Architext: An Introduction* (Berkeley: University of California Press, 1992).
11. Jeffrey Jerome Cohen, "Introduction: All Things," in *Animal, Vegetable, Mineral: Ethics and Objects* (Washington: Oliphaunt Books), 8.
12. Baz Kershaw, *Theatre Ecology* (New York: Cambridge University Press, 2007), 24.
13. Blair B. Kling, *The Blue Mutiny: The Indigo Disturbances in Bengal, 1859–1862* (Philadelphia: University of Pennsylvania Press, 1966), 16n3.

14. Ibid., 21.

15. David Arnold and Ramachandra Guha, eds., *Nature, Culture, Imperialism: Essays on the Environmental History of South Asia* (New Delhi: Oxford University Press, 1997); Elizabeth DeLoughrey and George B. Handley, eds., *Postcolonial Ecologies: Literatures of the Environment* (New York: Oxford University Press, 2011); Rob Nixon, *Slow Violence and the Environmentalism of the Poor* (Cambridge: Harvard University Press, 2013).

16. Jason Moore, "Ecology, Capital, and the Nature of our Times: Accumulation, and Crisis in the Capitalist World-Ecology," *American Sociological Association*, 18, no. 1 (2011): 107–46, 114.

17. For an early discussion of this growing topic, see Rohan Debroy, ed., "Nonhuman Empires," Special Issue of *Comparative Studies of South Asia, Africa, and the Middle East* 35, no. 1 (2015).

18. According to Nandi Bhatia, *Neel Darpan* "inaugurated the theater as a powerful weapon of resistance in the struggle for independence from colonial rule." *Acts of Authority / Acts of Resistance: Theater and Politics in Colonial and Postcolonial India* (Ann Arbor: University of Michigan Press, 2004), 22. For a discussion of *Neel Darpan* in the nationalist context, see also Sudipto Chatterjee, *The Colonial Staged: Theater in Colonial Calcutta* (Calcutta: Seagull, 2007). Ranajit Guha, however, underlines the play's truncated radicalism, noting that while it is critical of the planters, it espouses faith in the benevolence of the colonial state. Ranajit Guha, "*Neel-Darpan*: The Image of a Peasant Revolt in a Liberal Mirror," *The Journal of Peasant Studies* 2, no. 1 (October 1974): 1–46.

19. Michael Taussig, *What Color Is the Sacred?* (Chicago: University of Chicago Press, 2009), 154.

20. Kling, *Blue Mutiny*, 25

21. "Native Petitions Against the Indigo System in 1855," in *Strike, But Hear! Evidence Explanatory of the Indigo System in Lower Bengal*," compiled by Rev. J. Long (Calcutta: R. C. Lepage, 1861), 24–26.

22. Lalit Chandra Mitra, *History of Indigo Disturbance in Bengal: With a Full Report of the Nil Darpan Case Compiled by Lalit Chandra Mitra* (Calcutta: Provash Chandra Mitra, 1906), 37.

23. Ibid., 39.

24. It would be erroneous, however, to conclude that governmental intervention on behalf of the *ryots* was the main reason why the indigo industry in Bengal ceased to be profitable. By the 1860s, the indigo trade itself had become less lucrative. For the key features of the report, see *Minute by the Lieutenant-Governor of Bengal on the Report of the Indigo Commission Appointed Under Act XI of 1860* (Calcutta: Bengal Secretariat Office, 1861).

25. See Bhatia, *Acts of Authority/Acts of Resistance*, ch. 2. For the evolution of Bengali drama in the nineteenth century, see Chatterjee, *Colonial Staged*, and Prabhucharan Guha-Thakurta, *The Bengali Drama: Its Origin and Development* (Westport, Conn.: Greenwood Press, 1974).

26. At his trial, Long stated that by doing so, his intent was to forestall another full-scale rebellion. For details of Long's trial, see *Trial of the Rev. James Long for the Publication of The "Nil Durpun" with Documents Connected with Its Official Circulation* (London: James Ridgway, 1861).

27. In the original Bengali version (1860), the planter's name is P. P. Rogue, but, interestingly, in the English translation (1861), the planter is referred to as P. P. Rose.

28. Amiya Rao and B. G. Rao, *The Blue Devil: Indigo and Colonial Bengal, with an English Translation of Neel Darpan by Dinabandhu Mitra* (Delhi: Oxford University Press, 1992), 1.3, page 191. In reading the play, I have consulted the original Bengali edition published in 1860, the English edition translated in 1861 (at Long's behest), and the more contemporary edition translated and edited by Rao and Rao. All references to the play are to the latter edition (translated by Rao and Rao) and are hereafter cited parenthetically in the text by act and scene, followed by page number. The edition translated by Rao and Rao includes sections of the play omitted in the 1861 translation.

29. While dramatic performances were not censored when the play was published, lines such as these indicate the extent to which plays like *Neel Darpan* brought about the legislation of the Censorship Act in 1876, which "prohibited dramatic performances which [were] seditious or obscene, or otherwise prejudicial to the public interest." Bhatia, *Acts of Authority*, 19.

30. Rao, *Blue Devil*, 136.

31. Chatterjee, *Colonial Staged*, 225.

32. Bhatia, *Acts of Authority*, 41.

33. The idea of the less spectacular but equally potent nature of environmental violence is from Nixon, *passim*.

34. For a reading of the "rape scene," see Pamela Lothspeich, "Unspeakable Outrages and Unspeakable Defilements: Rape Narratives in the Literature of Colonial India," *Postcolonial Text* 3, no. 1 (2007): 1–9.

35. DeLoughrey and Handley, "Introduction: Toward an Aesthetics of the Earth," 10.

36. Ibid., 8.

37. Ibid., 13.

38. Curiously, Dutt, James Long's Bengali poet-translator (and a member of the Bengali literati), does not set these lines to rhyme, a further indication, perhaps, of their evocation of the influence of *jatra*. The changes and omissions in the 1861 translation deserve more sustained analysis and are beyond the scope of this essay.

39. Sumanta Banerjee, *The Parlour and the Streets: Elite and Popular Culture in Nineteenth Century Calcutta* (Calcutta: Seagull, 1989), 103.

40. Chatterjee, *Colonial Staged*, 124–25. For an insightful—and corrective—account of the intersection between "indigenous" and "Western" performance traditions in colonial Indian drama, see Aparna Bhargava

Dharwadker, *Theatres of Independence: Drama, Theory, and Urban Performance in India Since 1947* (Iowa City: University of Iowa Press, 2005), 144–46.

41. Kling, *Blue Mutiny*, 29.

42. Qtd. in Jeffrey Jerome Cohen, "Introduction: Ecology's Rainbow," in *Prismatic Ecology: Ecotheory Beyond Green*, ed. Jeffrey Jerome Cohen (Minneapolis: University of Minnesota Press, 2014): xv–xxxv, xi.

43. Kling, *Blue Mutiny*, 31.

44. Ibid.

45. The play on words here is obviously between a theatrical "actor" and the actor or actant in Latour's ANT discussed earlier.

46. Cleanth Brooks and Robert B. Heilman, *Understanding Drama: Twelve Plays* (New York: Henry Holt, 1945), 26.

47. For a reading of the consequent dialogism of drama, see Marvin Carlson, "Theater and Dialogism," in *Critical Theory and Performance*, edited by Janelle G. Reinelt and Joseph R. Roach (Ann Arbor: University of Michigan Press, 1992), 313–23.

48. Una Chaudhuri, *Staging Place: The Geography of Modern Drama* (Ann Arbor: University of Michigan Press, 1995), 80. For a suggestive account of how a metonymic—and unmetaphorized—reading of novelistic objects can vitalize them, see Elaine Freedgood, *The Ideas in Things: Fugitive Meaning in the Victorian Novel* (Chicago: University of Chicago Press, 2006).

49. Latour, "Agency at the Time of the Anthropocene," 11, original emphasis.

50. Ghosh, *Great Derangement*, 65.

51. Until 1967, the Flora and Fauna Act in Australia classified indigenous peoples as fauna, Deloughrey and Handley, "Introduction: Towards an Aesthetics of the Earth," 12. For a reading of the classification of indigenous tribes in colonial India, see Kavita Philip, *Civilizing Natures: Race, Resources, and Modernity in Colonial South India* (New Brunswick, N.J.: Rutgers University Press, 2003).

52. Sumanta Banerjee, *Parlour and the Streets*, 83, 94. In the play, the tropes from *jatra* are evident not only in the epilogue but also in the bawdy undertone to the conversation that characters such as Padi engage in.

53. Also, as Philip reminds us, it is not that indigenous knowledge-systems were not rational. *Civilizing Natures*, 16. Relatedly, for a cautionary note against overemphasizing the redemptive attributes of "folk theatre" in discussions of contemporary Indian drama, see Dharwadker, *Theatres of Independence*, 318. For a discussion of the orientalizing effects of incorporating folkloric forms into modern Indian theater, see Vasudha Dalmia, *Poetics, Plays, and Performances: The Politics of Modern Indian Theater* (Delhi: Oxford University Press, 2006), 199–212.

54. Moore, "Ecology," 108. As Moore puts it, "master processes," such as colonialism, are "at once a product[s] as well as producer[s] of far-flung and unruly relations between human and extra-human nature" (Ibid., 115). To bring Latour and Moore together within a span of a few pages, as I have done, may well seem odd given their vastly differing methodologies and the fact that Latour is interested in an ontological pluralism and Moore in a world-systems analysis. Nonetheless, if a "relational" view is key to Moore's analysis (see especially the chapter "Anthropocene or Capitalocene" in his *Capitalism in the Web of Life*), it is hard to think of how such a view might be fashioned without the vibrancy/materialism accorded to the nonhuman by Latour, Jane Bennett, and others.

55. Letter from A. Sconce to Secretary to the Government of Bengal, April 20, 1854. *East India Commission Papers*, printed by the order of the House of Commons, March 4, 1861. Compiled in Long, "*Strike, But Hear!*," 10.

56. Kling, *Blue Mutiny*, 31–32.

57. Tobias Menely and Margaret Ronda, "Red," in Cohen, ed. *Prismatic Ecology*, 30.

58. *Amrita Bazar Patrika*, December 12, 1872.

59. See Lisa Lowe, *Immigrant Acts: On Asian American Cultural Politics* (Durham: Duke University Press, 1996).

60. See Una Chaudhuri, "'There Must be a Lot of Fish in That Lake': Toward an Ecological Theater," *Theater* 25, no. 1 (Spring/Summer 1994): 1–23.

61. Alan R. Ackerman notes that "drama can[not] be understood apart from theatre. On the contrary, few would propose that we can make sense of drama without references to specific forms and venues of theatrical performance." In "The Prompter's Box: Toward a Close Reading of Modern Drama," *Modern Drama* 49, no. 1 (Spring 2006): 1–11, 3. With reference to Victorian drama, note Sharon Marcus's emphasis on "the phenomenology of its performance." "Victorian Theatrics: Response," *Victorian Studies* 54, no. 3 (Spring 2012): 438–50, 444.

62. Dana Luciano and Mel Y. Chen, "Has the Queer Ever Been Human?" *GLQ: A Journal of Lesbian and Gay Studies* 21, no. 2–3 (2015): 182–207.

63. Interview with Bruno Latour, "There Is No Earth Corresponding to the Globe," *Soziale Welt* 67, no. 3 (2016): 353–64, 357.

CHAPTER 2

Mourning Species
In Memoriam in an Age of Extinction

Jesse Oak Taylor

> Arise and fly
> The reeling Faun, the sensual feast;
> Move upward, working out the beast,
> And let the ape and tiger die.[1]
> —ALFRED TENNYSON, *In Memoriam*

These lines, from Alfred Tennyson's *In Memoriam* (1850), have haunted me for years. Read straightforwardly, the ape and tiger follow the reeling Faun and sensual feast as metaphors for the baser appetites and animal instincts, which must be left behind in the development of civilization. But read in the midst of the Sixth Extinction, when apes and tigers are, in fact, disappearing from the face of the Earth, along with thousands of other species, those metaphors cannot only be metaphors, but instead become synecdoches for the vanishing megafauna whose habitats are destroyed to sate the wants of that upwardly mobile species, *Homo sapiens*. The lines thus become essentially unreadable, not only because I struggle to face the prospect of extinction, but because what they mean to me seems diametrically opposed to what they presumably meant to their author. This is not merely an interpretive problem, but an ethical one. To "*let* the ape and tiger die" makes clear that if apes and tigers do go extinct, it will have been on my watch, that I will be culpable in their demise. Indeed, the poem seemingly *encourages* that outcome. To read *In Memoriam* in the Anthropocene is thus to encounter an ecological uncanny, in which the poem's familiar obsession with evolution, extinction, and deep time return in an unfamiliar guise.

This, in turn, presents a set of methodological problems about how to read in the Anthropocene. What kinds of evidence can a poem provide when encountered across the rupture marked by a new geologic age defined by human action? Dipesh Chakrabarty has argued that with the Anthropocene "the wall between human and natural history has been breached."[2] Human inscription is no longer purely the stuff of texts and images but has become legible in the geologic record as a constitutive feature of planetary processes. This means that the work of humanities no longer stops at the bounds of the human. The effort to affix the Global Boundary Stratotype Section and Point (GSSP), or "golden spike," that would make the Anthropocene designation official showcases this predicament in perhaps its most acute form, as stratigraphers explore possible "signatures" of human action legible within the geologic record.[3]

This essay explores how the signature of the Anthropocene might appear in literary history by revisiting the problem of extinction dramatized in *In Memoriam*, Alfred Tennyson's magisterial elegy for his beloved friend Arthur Henry Hallam, who died suddenly in 1833. *In Memoriam* was composed over seventeen years and finally published in 1850. Heavily influenced by Charles Lyell's *Principles of Geology* (1830–33), the poem offers extended, searching meditations on geologic time, evolution, and extinction a decade before Darwin's *Origin of Species* (1859). Thomas Henry Huxley described Tennyson as "the first poet since Lucretius who has understood the drift of science."[4] The poet's wide reading in astronomy, geology, natural history, physics, and theology leave their footprints everywhere in the text, the subject matter of which extends from the workings of the poet's own mind to the ends of the universe. *In Memoriam* is an important work of Victorian evolutionary theory in its own right, one that profoundly influenced the Victorians' sense of themselves as geologic agents and popularized ideas ranging from deep time to thermodynamics, the nebular hypothesis to extinction. Tennyson's poem takes on new resonance in keeping with Wai Chee Dimock's use of that term, as it echoes in our own encounter with versions of many of these same ideas.[5] Like the Victorians, we struggle to account for human history in geologic time, to question what it means to live—and die—as one species among many.

In the remainder of this essay, I will ask a series of questions of Tennyson's poem in order to dramatize the unsettling, retrospective work of Anthropocene literary history. First, how does the poem's invocation of extinction shift when it is read (as it was indeed written) in the midst of a mass extinction event caused by human action? That is to say, what happens when Tennyson's treatment of extinction is viewed not as the inevitable work of a careless and rapacious Nature, but as the historical work of a careless and rapacious "Anthropos"? Second, does his treatment of *human* evolution take on new valence when the human is viewed as a distinctive *kind* of species operating as geohistorical force within the Earth system? Chakrabarty argues that "we humans never experience ourselves as a species," precisely because "one never experiences being a concept."[6] A similar problem arises in relation to the question of what is lost when a species goes extinct because, by contrast with an individual, a species *isn't alive*. Or, more accurately, it lives only on a second-order plane irreducible to the life displayed by any one instantiation of it. Construing species as *mournable*, then, depends on reconceiving elegy, the poetic form that enables and formalizes mourning, not as a lament for an individual, but for an abstraction. While mourning may not appear politically efficacious, it is integral to ascribing to the species ethical and political value distinct from that of the individual animal or person, thus rendering them worthy of protection.[7] Furthermore, the act of commemoration itself grants the departed entity a form of ongoing agency, an ability, however ghostly, to shape the future.

In Memoriam is a fruitful location to take up this inquiry in part because it has been so influential in cementing "Nature, red in tooth and claw" (56.15) as the definitive agent of evolutionary ecology. Many of Tennyson's readers have focused on the poem's attempt to grapple with the transience of the world and the place of humanity in it, such that Hallam's death and the poet's grief are enfolded within the inevitable annihilation of the human species. For example, David Shaw sees the following lines as a "time-lapse photograph of the earth extending over billions of years":

> There rolls the deep where grew the tree,
>   O earth, what changes has thou seen!
>   There where the long street roars, hath been
> The stillness of the central sea.
>
> The hills are shadows, and they flow
>   From form to form, and nothing stands;
>   They melt like mist, the solid lands,
> Like clouds they shape themselves and go. (123.1–8)

Distinguishing *In Memoriam* from the elegiac tradition in which the physical world mourns along with the poet, Shaw argues that Tennyson's poem depicts "a savage world that is still strangely evanescent," a world in which "the death of Tennyson's friend may be of the same order as the melting of an icicle."⁸ Similarly, Eleanor Bustin Mattes suggests that Tennyson's lines expressing the fear that man and all his works might "Be blown about the desert dust, / Or seal'd within the iron hills?" (56.19–21) "echo Lyell's conclusion, 'that none of the works of a mortal being can be eternal.'"⁹ However, the Anthropocene presents a paradoxical inversion of this predicament because it is the *inscription* of the human rather than its erasure that has become the source of anxiety. As Bronislaw Szersynski puts it, "the fate of 'man' in the Anthropocene is not that he will be erased, but that he will be made immortal, as a trace preserved forever in the rock."¹⁰ The Anthropocene thus takes shape as an age of prophetic elegy, when to imagine the future is to imagine not so much the erasure of the present as the legible traces of that obliteration, to view the objects of the present world as fossils-to-come.

Noting this, in turn, helps re-think the work of elegy in an age of mass extinction. Elegy is at once the paradigmatic genre of the environmental imagination, and one of which ecocritics have become increasingly skeptical. Margaret Ronda notes that elegy's traditional reliance on natural imagery to provide "symbolic correlatives for loss and consolation" means that the "end of nature" is both an elegiac idea and a condition that renders elegy itself impossible, such that the poet is left, suspended in melancholia, in effect, mourning *for* mourning. Thus, Ronda asks, "Can there be elegy . . . without the 'absolute other' of nature?"¹¹ Such melancholic suspension is in keeping with Jahan Ramazani's argument that "modern elegy" more broadly departs from the genre's emphasis on closure working instead to "resist consolation," "sustain anger," and "reopen the wounds of loss."¹² Similarly, Timothy Morton suggests that ecological elegy too often "presupposes the very loss it wants to prevent," and must instead "mobilize some kind of choke or shudder in the reader that causes the environmental loss to stick in her throat, undigested. . . . [refusing] to work through mourning to the (illusory) other side."¹³ Ursula Heise acknowledges melancholy to be "an integral part of the environmentalist worldview," that can help ascribe political and ethical value to nature, but ultimately she, too, sees it as a dead end, suggesting that a shift from elegy to comedy might "enable the imagination not so much of the end of nature as its future."¹⁴ Each of these cases focuses on the challenges of writing in response to avowed and recognized ecological crisis. However, the predicament raised by *In Memoriam* is different, in part because the poem was written before the Anthropocene came into focus as such. Tennyson's poem stages ecological disruption as a theater for the poet's grief, presenting as fantasy the loss that subsequent ecopoets lament. In this sense, *In*

Memoriam seems to fit the familiar narrative of unintended consequences, in which ecological awareness comes too late, such that all we can do in retrospect is mourn. However, Christophe Bonneuil and Jean-Baptiste Fressoz counter this notion that the Anthropocene came upon us unawares, noting that global ecological crisis was theorized, and the historical processes undergirding it contested, at every turn. Thus, they argue: "The history of the Anthropocene is not one of a frenetic modernism that transforms the world while ignorant of nature, but rather of the scientific and political production of a modernizing unconscious."[15] The poem's uncanny echoes thus reverberate from within the modernizing unconscious, issuing unsettling reminders that we knew what we were doing all along. With this in mind, I want to reject the nominally historicist notion that Tennyson's ape and tiger can be read in purely figurative terms, innocent of the actual animals' demise. *In Memoriam* emerged from, and participated in, historical forces that were catastrophic for the world's species. The Victorians were acutely aware of themselves as agents of extinction, and the poem itself thematizes that awareness, putting a new gloss on elegy's haunting power, and the chance it offers to commune with the dead.

Memorializing the dead hinges on what Robert Pogue Harrison calls "the afterlife of the image."[16] According to Harrison, "We dispose of what is inanimate in the dead so that they may find their way into realms of the spirit—realms to which the living, by virtue of their existential self-overreaching, have native access. To be mortal means to *be* the place of this imaginary afterlife."[17] Memory extends the "realms of the spirit" into the realms of the living, but it relies on physical, poetic, or otherwise external commemoration as a supplement to that effort. Extending this argument into evolutionary biology, Eduardo Kohn writes that "all of life . . . houses by virtue of these constitutive absences, the traces of all that has come before it—the traces of that which it is not."[18] Writing in a similar vein in the introduction to their magnificent collection *Arts of Living on a Damaged Planet: Ghosts and Monsters of the Anthropocene,* Elaine Gan, Nils Bubant, Anna Lowenhaupt Tsing, and Heather Swanson invite us to wander "through landscapes, where assemblages of the dead gather together with the living" because "in their juxtapositions, we see livability anew." "Ghosts," here, include not only those vanished forms that can be aesthetically traced, but also those that are preserved as palpable absences in the existence of the species with which they co-evolved and have left behind. Thus, "trees that grow back when cut down, such as oaks, may have evolved that ability in times when elephants trampled them. The ghosts of lost animals haunt these plants, even as the plants live on as our companions in the present."

Because evolution is always, to some extent, co-evolution, "extinction is a multispecies event."[19] So, too, are mourning and conservation. Being species means that we retain the remains of other species deep within our being: our biology, our language, our myths, and our poetry. Recovering these persistent traces is the task of elegy in an age of extinction. Diana Fuss describes elegies as "voices without bodies," which is to say that elegy attempts to recover for the written word a form of immediacy and presence that precedes writing and would be possible only when speaking in the living presence of the deceased.[20] Elegy thus entails a form of magic, a conjuring back to life, in which the poet is not simply haunted by, but becomes the prosthetic voice of, the departed. Devin Griffiths sees precisely this turn at work in *In Memoriam*, which he argues is not simply a *lament* for Hallam, but coauthored by him, insofar as Tennyson not only writes *about* Hallam but also under his influence, ventriloquizing aspects of his dead friend's poetic style. Hallam continues to live within the poem, which "has to remain open to his intercession, his ability to reach through time and shape its movement."[21] Because it enables the dead to speak, elegy forces us to remain open to their intercession. This, in turn, is why elegy remains a vital ethical and political project for the Anthropocene. In coping with the losses marked by extinction, we must find ways to recover that portion of a species that exceeds the body, and to think seriously about the agency such undead creatures continue to display in the world. Griffiths's emphasis on *In Memoriam* as a collaborative work also extends to testimony of extinct species as advocates, interceding on behalf of the creatures and biomes that remain.

Identifying the *species* as a locus of value that does not collapse back into the individual is both a key aspect of the work of elegy in the Anthropocene and a problem that *In Memoriam* takes up directly. The poem addresses the problem (and power) of abstraction on multiple registers. As Timothy Pelatson notes, *In Memoriam* "repeatedly offers itself as a representative unit of human history, a model, for better or for worse, of the individual life and the life of the species." This arises in part because of the poem's formal structure: "a long poem made up of short poems, *In Memoriam* naturally interests itself in the way that short poems build into longer ones, the relations of part to part and part to whole."[22] Thus, the poem formalizes the processes of aggregation and abstraction inherent in the species concept. *In Memoriam* consists of 133 cantos of varying length, all utilizing the same *abba* rhyme scheme, with the two middle lines enjambed, a distinctive form now known as the *In Memoriam* stanza. This formal unity would seem to belie Tennyson's claim that he "did not write them with any view of weaving them into a whole, or for publication, until I found that I had written so many."²³ Meanwhile, that claim itself suggests that the poem's recurring form arose first not as an organizing principle but, rather, as a mode of self-defense, an attempt to build a bulwark against the waves of grief. The stanza form also provides a formal engagement with the questions of geological change and evolution that run through the poem.

The stanzas are laid down like the strata of the fossil record, preserving within their interlocking layers the remnants of the lost. They are also *iterative*, allowing new forms, patterns, and species of meaning to emerge in the passage from one stanza to the next, providing an innovative (if inadvertent) solution to one of the challenges of narrating evolutionary change: namely, the fact that narrative is inherently teleological, linking events in a logic of cause and effect, while evolution is random and contingent, arising out of iterative processes with no guarantee of linear progress (though many Victorians distorted evolution into an inherently progressive doctrine). Hence, I say "inadvertent" both because Tennyson had adopted the form prior to his reading of Lyell and because the problem it solves is not one that he would have attributed to evolutionary theory. Like many of the observations I will make in the text that follows, this solution only makes sense after the fact, suggesting that the true significance of the poem becomes visible only in retrospect, in a literary-historical version of evolutionary retrospective causality, in which purpose and utility are always belated.

In this way, the Anthropocene appears within *In Memoriam* as a kind of unconformity, a fold in literary history. As Eric Gidal explains, unconformities are disruptions within Huttonian geology and the stratigraphic method upon which Anthropocene dating depends, wherein deeper strata are older, thus enabling a cross-section or core sample to be read as a quasi-linear account of history. Unconformities, however, mark the points where that logic is disrupted, when "compressions of distinct eras of sedimentation caused by orogenic shifts and subsequent erosions" produce "radically discordant appearances in the layering, or 'superposition' of strata." As such, unconformities offer "physical manifestations of heterogeneous time." Gidal builds on this concept to argue for what he calls *biblio-stratigraphy*, in which the work of reading entails "tracing the signatures of social and spatial changes" in order to perceive the "dynamic and protean nature of environmental and social conditions over wide scales of time."²⁴ This concept is productive as a model for literary-historical analysis precisely because, as Gidal shows, it does not depend on recovering an "ethic of stewardship" or a "mandate for ecological justice" (common ecocritical preoccupations) out of the artifacts of the past, but rather traces the process by which such artifacts can "*become* important environmental records for our own moment" and "offer witness to rapid and precipitous change."[25] My contention is similar, insofar as I am not arguing, in biographical terms, that Tennyson was an advocate for endangered species, or even that *In Memoriam* intimates a logic of inter-species care, but rather suggesting that his meditations on extinction can now be called on as a form of trans-historical witnessing that allows *us* to extend our awareness beyond the limits of an individual life. Writing in the ferment of industrialization, empire, and evolutionary theory that we can now recognize as the dawn of the Anthropocene, Tennyson produced a poem in which we can now trace the lineaments of the epoch's emergence.

In Memoriam thus becomes an elegy *for* the Anthropocene precisely insofar as it is a work bequeathed to the future, through which we can glean some meaning in our catastrophic present by imagining a future retrospect. The Anthropocene concept itself depends on precisely this move, recasting the present as history, encoded in a stratigraphically legible trace at some far distant future date. Anthropocene discourse is thus at once future-oriented and inherently retrospective. This, too, accords with both Tennyson's normal manner of work (he often composed the ends of his poems first) and his preoccupation with retrospection. As Herbert Tucker describes: "Even when . . . the process of composition was straightforward, Tennyson delighted to enclose his poems in frames that give the illusion of retrospective return."[26] Griffiths argues that *In Memoriam* "weaves a comparative impulse" into its form because "stanzas continually fall back, review, and revise their own understanding."[27] Retrospection and *telos* come into tension, as Tennyson ultimately affirms a teleological view of progress, drawing solace from "one far-off divine event,/To which all creation moves," but the iterative form makes clear that this imposition of purpose is a conscious choice rather than poetic (or natural) imperative (Epilogue 143–44). As many critics have observed, *In Memoriam* could easily go on forever. As Griffiths argues, "the poem accrues as a web of interconnected experiences, a mosaic of encounters, a pattern of diffraction, rather than as a stately narrative that moves its subject through coherent stages of grief and acceptance."[28] Hence, any narrative discernible within *In Memoriam* must be superimposed upon its undergirding strata much the way myths have long ascribed purpose to geological phenomena, which then become inscribed as stories in the repositories of cultural memory.

In Memoriam is filled with language that slides among geological, biological, and literary form. As such, form emerges from the poem as a category in which those multiple registers become commensurate. That commensurability, in turn, offers an opportunity to extend the links that Caroline Levine establishes between aesthetic and social form beyond the human.[29] Consider the poem's most explicit mention of extinction:

> "So careful of the type?" but no.
>   From scarped cliff and quarried stone
>   She cries, "A thousand types are gone;
> I care for nothing, all shall go." (56.1–4)

Here, "type" refers to species as opposed to the individual, thus revising the previous canto's reflection, "so careful of the type she seems / So careless of the single life" (55.7–8). However, a more profound shift occurs when we move from the death of the individual to that of the species. A "type" is a category, an abstraction to which any number of individuals might be aggregated based on some shared characteristic or characteristics, while ignoring the particularities that would distinguish between them. "Type" is also printer's type: a material object by which an abstract sign, the *letterform*, becomes a legible trace imprinted on a page. It thus physically operates in a manner analogous to a dinosaur's foot leaving an imprint in sand that might one day be compacted to stone such that it becomes a legible trace millions of years later. The fossil record is composed of "types"—not just in the sense of categories, but also in the imprints of form. The obvious objection to this line of thinking is that the individual letterform isn't alive, whereas the dinosaur is—or was. But the individual dinosaur also isn't a species. Indeed, in order to *become* a species, an individual has to be abstracted into an image, or type. As Alex McCauley has argued, in the nineteenth century this translation from animal into media, via taxidermy, was an explicitly embodied, and thus messy, smelly, violent process of being shot, skinned, boiled, preserved, and stitched back together, shipped around the world and put on display.[30] In order for species to take shape, animals had to become specimens. The "type" had to be abstracted from the individual life, and converted physically, violently, into a sign. The modern scientific idea of species was thus violently wrested from the world, abstracted from the blood and viscera of thousands of animal bodies. The same was true of plants. As Jim Endersby explains, the illustrations in botanical catalogs were "composite images, drawn from numerous different sketches and specimens," that nonetheless came to define the "type" to which all subsequent specimens would be compared in the fraught debates between the "lumpers" and "splitters" over species classification.[31] Darwin himself uses "form" and "species" interchangeably throughout the *Origin*, as when he writes that "all living and extinct forms can be grouped together in one great system," reflects that "the more ancient any form is, the more ... it differs from any living form," or describes a fossil find by noting that "only one or two species are lost forms and only one or two are new forms."[32] Indeed, he actually seems to prefer "form" to "species," at least when referring to any empirical observation or evidence, perhaps because "form" can refer to either the species as a whole or the properties of the individual organism, whereas "species" refers *only* to the abstraction or aggregate, and thus doesn't actually appear in the world as such. Put differently, because the "struggle for existence" occurs in the living world, species must be instantiated as forms in order to act, interact, and reproduce. For Darwin, then, form cleaves body and image, individual and species. Insofar as it negotiates the formal relationship between dead bodies and deathless idea, we might even think of *The Origin of Species* as itself a kind of elegy, in which death is transmogrified into meaning via the magical power of language.

There is no composition without decomposition. Elegy is thus not simply a poetic mode concerned with death and mourning, but in some respects the paradigm of poesis itself, insofar as what we call "making" really means repurposing and reshaping existing matter, translating the messy, evermutating world of life and death into the more durable, if abstract, symbolic domain, much the way that Derrida reflects that all writing is designed to outlast its author and thus carries the implicit promise of death.[33] Like images on a cave wall, forms can endure even millennia after the bodies they copied have been re-metabolized. John MacNeill Miller has recently argued that *In Memoriam* performs precisely such a conjuring trick upon Hallam's material remains, which were in an advanced state of decay by the time they were returned to England from Italy to be buried. As Miller notes, the poem is "riddled with recurring terms such as must, mould, rank, and blow, words whose buried secondary meanings all involve decomposition." Thus, Miller suggests, *In Memoriam* "shares the tomb's dubious function of shoring up the boundaries of a body whose distinction from its material environment wanes with each passing day."[34] In order to transform Hallam into a deathless figure, enduring voice, and even (in Griffiths's reading) active co-author, in other words, the poem must first do away with "a messy collection of organic material steadily advancing through the process of rotting." As Miller explains, "in obscuring rot, we obscure ecological realities, turning our backs on a vital connection between individual human beings and the broader biotic community of which we are a part."[35] The question, then, becomes how we might rethink elegy in the context of ecological processes of renewal, which are predicated not on a paradoxical fantasy of bodily cohesion but rather on the ongoing metabolism of living systems—of recognizing with Donna Haraway that we are "compost"—without losing sight of the uniquely animating qualities of form or minimizing the trans-historical community afforded by the symbolic realm.[36]

This, in turn, brings me to the third valence of "type" operative in *In Memoriam*, namely Biblical typology, in which the type both forecasts and is superseded by its antitype, a connotation that Tennyson invokes explicitly and which links literary history with the inherently retrospective explanatory rubric of evolutionary biology. In this context, typology becomes useful precisely insofar as it hinges on *retrospective* reading. As such, typology actually aligns with a key aspect of evolutionary theory, by which purpose and utility always come *after* the mutations that enable them. The implications of this shift become particularly evident in Canto 118, which traces the geological history of the planet, beginning with the injunction to "Contemplate all this work of Time/The giant laboring in his youth," before turning to human evolution:

> till at last arose the man:
>
> Who throve and branch'd from clime to clime,
> > The herald of a higher race,
> > And of himself in higher place,
> If so he type this work of time
>
> Within himself, from more to more; (118.9–17)

"Time" is first introduced as an agent of geological change, only to be superseded by "man" who comes to "type this work of time, within himself" and in so doing forecasts the coming of a "higher race." For Tennyson, "higher race" and "higher place" operate within a typological framework as references to a religious afterlife.[37] However, that structure is here applied to evolution in ways that now appear symptomatic of the emergence of the Anthropocene. To "type the work of time" is to adopt, take over, or subsume the work hitherto carried out by Earth systems, a transition in the "work" of the Earth system that aligns with recent accounts of the Anthropocene, including not only Paul Crutzen and Eugene Stoermer's initial proposal dating the epoch to 1784, but also Simon Lewis and Mark Maslin's "Orbis Hypothesis." Lewis and Maslin propose that the Anthropocene be dated to 1610, and the aftermath of European conquest in the Americas, which resulted in "a swift, ongoing, radical reorganization of life on Earth without geological precedent," coupled with the deaths of some 50 million

Native Americans due to war, enslavement, starvation, and disease, a dying so great that it remains legible in the polar icecaps.[38] This genocide also lies hidden within *In Memoriam*, echoing in the imperialist undertone of "higher race."

The process by which "man" "throve and branch'd from clime to clime" is one that Lyell (and many others) linked explicitly to extinction. In *Principles of Geology*, he writes, "if we wield the sword of extermination as we advance, we have no reason to repine the havoc committed" because "every species which has spread itself . . . over a wide area, must, in like manner have marked its progress by the diminution, or entire extirpation, of some other."[39] Lyell's complacency is emblematic of the "extinction discourse" that Patrick Brantlinger traces across the political and disciplinary spectrum of nineteenth-century thought, which framed the extinction of "savage" or "primitive" peoples as inevitable, often through a form of "proleptic elegy" that "mourns the lost object before it is completely lost."[40] While Brantlinger makes very little of the overlap between the extinction of "primitive" races on the one hand and wild animals on the other (he never mentions endangered species, and seems barely aware of animals, habitats, or wilderness *except* as discursive formations), his argument dovetails neatly with Morton's idea that ecological elegy "presupposes the very loss it wants to prevent."[41] In each case, proleptic elegy's future-perfect orientation is understood to facilitate loss, rather than forestalling it. It may come as a surprise, then, to learn that Brantlinger sees *In Memoriam* pushing *against* the broad current of extinction discourse.[42] Brantlinger points to the lines in which man becomes

> A monster then, a dream,
> A discord. Dragons of the prime,
> That tare each other in their slime,
> Were mellow music matched with him. (912)

For Brantlinger, this becomes evidence that *In Memoriam* offers up "another version of *Jurassic Park*, only this time the dinosaurs have not returned; they have been replaced by that greater monstrosity, mankind."[43] However, he turns from this observation to Tennyson's anxiety over human extinction, thus seemingly abandoning the insight opened by his *Jurassic Park* reference, namely that the real monstrosity is human history and the violence that attends it, including the fact (acknowledged explicitly by Lyell) that *Homo sapiens*' arrival in the geological record inaugurates an age of extinction.[44] These lines reflect the anxiety of guilt rather than apprehension of demise. It is the human as *agent* of extinction, as well as victim of it, that incites Tennyson's concern. Anthropogenic extinction is not natural but monstrous, a "discord" exceeding even its precedents in the compounded death of the fossil record.⁴⁵

In Tennyson's agonized reflections, then, we can begin to see an incipient awareness of humanity's role as an agent within geophysical processes, which is to say not simply the evolution of *Homo sapiens*, but rather the emergence of the Anthropos as a distinctive force within the Earth system. This may seem precisely the point at which I am pushing Tennyson farther than he could possibly have been prepared to go, except that in the very next lines he turns from conquest to industrialization, and thus from one account of the Anthropocene (1610) to another (1784). The reflection that man might "type the work of time/within himself" is followed by an extended metaphor linking "life" to industrial processes:

> life is not as idle ore,
>
> But iron dug from central gloom,
>> And heated hot with burning fears,
>> And dipt in baths of hissing tears,
> And batter'd with the shocks of doom
>
> To shape and use. (118.18–25)

Here, extractive industrial capitalism, which mines ore from the ground, smelts it in great furnaces, heats, forges, and tempers it "to shape and use" becomes the model for efficacious life, such that life itself becomes a question of use value. These industrial metaphors would not have been lost on Tennyson's Victorian readers. T. H. Huxley (an admiring reader of Tennyson) makes precisely the same move in his lecture "On the Formation of Coal" (1870), which imagines club-moss being compacted into coal beds, where they lay for millennia until the arrival of James Watt in the late eighteenth century: "The brain of that man was the spore out of which developed the modern steam-engine, and all the prodigious trees and branches of modern industry . . . coal is as much an essential condition of this growth and development as carbonic acid is for that of a club moss."⁴⁶ Tennyson was thus participating in a widespread Victorian discourse that situated industrial production against the backdrop of deep time and the natural history of coal. Indeed, science writer Arabella Buckley, whose books sought to introduce children to the wonders of science, would later quote Tennyson to similar effect. Reflecting on the usefulness the lives of prehistoric plants acquired when combusted in Victorian steam engines, she reminds her readers:

> That nothing walks with aimless feet,
>   That not one life shall be destroyed,
>   Or cast as rubbish to the void,
> When God hath made the pile complete. (54.5–8)[47]

Buckley not only draws on Tennyson's assurance that industrialization provided a model for the "usefulness" of life, but also, more vertiginously, invites her young audience to take comfort in the anonymity of their own lives when considered against the "void" of deep time by virtue of "the sunbeams which those plants wove into their lives" (192). In each case, the depiction of fossil energy entwines it with the geological history of the planet such that industrialization is presented not as a *rupture* with natural processes or evolution but rather as an extension, concentration, and acceleration of them. Industrialization marks a phase transition in the Earth system, one that comes "dipt in baths of hissing tears" and echoes with the "shocks of doom." In so doing, it not only alters the operations of planetary processes but also imbues them with purpose, thereby rendering those material processes open to interpretation. This, as I see it, is the real logic behind the analogies that Tennyson, Huxley, and Buckley all draw between industrialization and the Earth system, which serve not so much to naturalize industrialization as to *de*naturalize geology. Reading *In Memoriam* in the Anthropocene means not only situating the poem against the backdrop of Earth's history but also, far less intuitively, extending poetic interpretation back into the geologic record.

Confronting such deep-temporal conjuring tricks lies at the heart of thinking about agency and species being in the Anthropocene. The "work of Time" will now be performed by "man," but only once man is construed as an industrial agent. Extractive industry not only converts *material* from "idle ore" into iron that has been mined, smelted, forged, and tempered in order to be shaped to use; it also, Tennyson suggests, imbues "life" with purpose, a purpose that thus becomes inextricable from the arts of industry. "Man" not only "types the work of time" but also becomes the "herald of a higher race," that scaled-up version of the human operating as a force of nature and seemingly oblivious to the suffering and death it wreaks upon the world—a voice that might well proclaim "a thousand types are gone;/I care for nothing, all shall go." This is, in other words, the voice of the much-maligned, much-debated "Anthropos," the agent at the heart of the Anthropocene.

To use an admittedly anachronistic term, I am suggesting that Tennyson is theorizing the emergence of the Anthropos as a kind of posthuman cyborg—a technologically mediated, imbricated, and enhanced (or at least magnified) being, in this case by industrialization, capable of "typing" the work of Time within himself, and thus of inscribing himself (or itself) within the geologic record. The "higher race" is not the human per se but rather a newly industrialized assemblage or emergent phenomenon as understood in systems theory—a "higher order" phenomenon irreducible to the sum of its parts. "Man," the nominal subject of Enlightenment humanism, thus becomes the herald of that higher race, as in the one who announces and enables its arrival. This would fit with Jason Moore's argument that Cartesian dualism, understood broadly as the separation of "nature" and "society" within Enlightenment science, undergirds the rapacious appropriation at the core of what he prefers to call the *Capitalocene*.[48] Many others have joined in this critique, noting that it is not *humans* but rather specific historical, economic, and technological processes that are disrupting Earth systems, and that furthermore those processes (capitalism and empire most notably) are themselves predicated on exacerbating and exploiting inequalities among humans. It is not my intention to minimize inequality as a constitutive force within the Anthropocene, but rather to note, following Chakrabarty, that human impact on the Earth system—even if that action is practiced by a small subset of actual humans—nonetheless marks the emergence of a new mode of human being, one that "has no ontology" and lies "beyond biology," and hence "acts as a limit" to any ontological account of the human.[49] Furthermore, I want to suggest that Tennyson's shift to typology is symptomatic of that new mode of being—the human as planetary force—that is of an entirely different precisely because of the introduction of industrial technology and fossil energy, but that nonetheless arises out of integral preconditions that form the basis of that transition.

Crucially, industrialization and conquest register in the poem not as literal force, but rather as extended metaphors, a turn to the figurative that is both appropriate for a mode of being that has "no ontology" *and* makes recourse to the symbolic realm, that aspect of human thought which most distinguishes it from that of other animals.[50] Victorian evolutionary theory was plagued by a similar double movement as it sought to understand humans as at once one species among many and a species of a different order, a self-reflexivity that we can now perceive as the dawning self-awareness of a planetary force. Tellingly, it is Canto 118 that ends with the lines, quoted at the beginning of this essay, that outline the casualties of the Anthropocene's emergence all too clearly: "move upward, working out the beast,/And let the ape and tiger die."

In this regard, *In Memoriam* becomes a threshold text, both marking the point at which it stops being possible to think the human as other than one among innumerable evolved and evolving species, and yet equally impossible to avoid the calamitous impacts that human history has wrought within the web of life. The impact of the human species within the Earth system is coeval with the understanding of human evolution itself, an alignment that I treat not as coincidence but as part of a broader pattern of synchronization between the material processes at the heart of the Anthropocene and the ideas (and modeling technologies) that render them legible.[51] In this view, *In Memoriam* becomes an elegy not only for the myriad species vanishing before they can even be named, but also the human itself, for the conception of "Man" that has become obsolete, and of which Hallam comes to signify the last best instance. The human that is lost is the human as other-than-Anthropos; the human that had yet to become a catastrophe for life on Earth. Tennyson's fear at being obliterated within the geologic immensity of time has shifted to the signature of the Anthropos that invokes annihilation in the very moment it is inscribed not because it *won't* endure, but because it will. In accounting for such retroactive shifts in meaning, Anthropocene reading demands attending to texts' afterlives (and perhaps their pre-lives), to the weird historical coincidences that attend their composition, and to their ever-emergent futurity.

In the final section of the poem, Tennyson asserts his belief in divinely ordained progress: "One God, one law, one element, / And one far-off divine event, / To which the whole creation moves" (Epilogue 142–44). This vision is preceded a few lines earlier by a vision of future generations that echoes his earlier conception of a "higher race" construed through its mastery of Nature. Thinking of his sister's children, Tennyson describes them as:

> a closer link
> Betwixt us and the crowning race
>
> Of those, eye to eye, shall look
>   On knowledge; under whose command
>   Is Earth and Earth's, and in their hand
> Is Nature like an open book; (Epilogue 125–32)

In light of an ever-expanding human population, a poem that concludes by placing its faith in the next generation may seem profoundly counterintuitive, even naïve. Rather than marking the point of human control "under whose command / Is Earth and Earth's," the Anthropocene buffets us with the "shocks of doom": superstorms, antibiotic resistance, ecosystem collapse, mass extinction, disappearing islands, and millions of refugees on the move. Nonetheless, I can no longer read the preceding lines without thinking of Stuart Brand's "eco-pragmatist" prognosis for the Anthropocene: "we are as gods, and we must get good at it."[52] Indeed, the leading candidate for the "golden spike" now appears to be neither conquest nor industrialization, but 1945, a century after the publication of *In Memoriam*, when the first nuclear tests align with the beginning of the "Great Acceleration" in fossil fuel use and population growth that continues unabated.[53] By this measure, Tennyson's "crowning race" finds voice in Robert Oppenheimer's reflection on the Trinity Test: "I am become death, the destroyer of worlds." Whether it is dated to the genocidal conquest of the Americas, the shift to fossil fuel, or the nuclear bomb, the Anthropocene is an epoch writ in death, which is part of why elegy may be its definitive poetic mode. However, the GSSP will mark not an end, but a beginning. As Jeremy Davies notes, the Anthropocene will only truly begin when the Earth system achieves a new era of stabilization. Until then, we are living in the boundary event that marks the end of the Holocene.[54] Affixing the GSSP and naming the Anthropocene redefines the history that precedes it, but it does not predetermine the future. The end of the Anthropocene remains as much a "far off event" for us as it was for Tennyson. It is a story yet unwritten.

In Memoriam's engagement with typology, geologic time, and extinction have all been well established, but the meaning of these familiar features changes when they are read as symptoms of the emergence of the Anthropocene, or premonitions of its arrival. Tennyson's poem is an elegy for the Anthropocene, not only because it speaks to so many themes of our geologic contemporaneity—evolution, extinction, geologic time—but also, paradoxically, because of the very historical gap that makes the Anthropocene something that *In Memoriam* shouldn't, properly speaking, be about. Whatever the Anthropocene is and whenever it started, it isn't over and none of us will see the end of it. Any formulation of the concept is prospective. Thinking about the Anthropocene is an exercise in imagining a future retrospect: how the age of humans will have become legible within the geologic record for millennia to come. Historical artifacts, read retrospectively, may in fact be the only way of rendering the Anthropocene legible. *In Memoriam* is about the Anthropocene not in spite of the fact that it was written before the Anthropocene concept came fully into view, but because of it.

Treating *In Memoriam* as an elegy for the Anthropocene expands upon the stratigraphic search for the "signature" of the Anthropos because an elegy is not simply a record, or even a memory, but a commemoration, a memorial and a conjuring back to life. Fuss argues that "enfolding the dead in its lyrical embrace, *In Memoriam* . . . shows not just how elegy might be ethical but how ethics might be elegiac."[55] Elegy draws its greatest force from the inevitability of death and the impossibility of that which it seeks to invoke: the absent voice. As such, it offers a model for thinking about ethics and endurance in the face of impossibility, whether the limits of our comprehension or the sheer scale of ecological catastrophe. Elegy is one of the oldest poetic forms precisely because it wrestles with the ultimate irreconcilable fact of life. Part of the shift from grief to commemoration is that commemoration is a public act, as state funerals, war memorials, and other instances of public mourning make explicit. Indeed, the publication of *In Memoriam* secured Tennyson the position of Poet Laureate, in which capacity he was frequently asked to compose epitaphs and memorials for public figures. This public function is vital in the present context because, as Jedediah Purdy has argued, "no 'we' that could grapple with the crises of the Anthropocene exists yet."[56] Elegy can help bring that "we" into being, channeling the community that arises in shared loss into new polities, new solidarities, and new forms of democracy. *In Memoriam*, like the Anthropocene, is a plea for future redress, a call awaiting an answer.

Notes

An earlier version of this chapter appeared in *Victorian Studies* 58, no. 2 (Winter 2016), and is reproduced with permission from Indiana University Press. I would like to thank Cornelia Pearsall, Devin Griffiths, Nathan K. Hensley, and Philip Steer for their generous (and generative) comments as I worked to revise and expand it here.

1. Alfred Tennyson, *In Memoriam* (1850), edited by Erik Gray (New York: Norton, 2004), 118.25–28. Hereafter cited parenthetically in the text by canto and line number.

2. Dipesh Chakrabarty, "The Climate of History: Four Theses." *Critical Inquiry* 35, no. 2 (2009): 197–222, 221.

3. Colin N. Waters, et al., "The Anthropocene Is Functionally and Stratigraphically Distinct from the Holocene," *Science* 351.6269 (2016): aad2622. For discussion of the "signature" of the Anthropocene as a problem for reading, see Tobias Menely and Jesse Oak Taylor, Introduction to *Anthropocene Reading: Literary History in Geologic Times* (State College: Penn State University Press, 2017).

4. Quoted in John Holmes, "'The Poet of Science': How Scientists Read Their Tennyson," *Victorian Studies* 54, no. 4 (Summer 2012): 655–78, 656.

5. Wai Chee Dimock, "A Theory of Resonance," *PMLA* 112, no. 5 (1997): 1060–71.

6. Chakrabarty, "Climate of History," 220.

7. Ursula K. Heise, *Imagining Extinction: The Cultural Meanings of Endangered Species* (Chicago: University of Chicago Press, 2016), 34–35. On the political efficacy of mourning, see Judith Butler, *Precarious Life: The Powers of Mourning and Violence* (London: Verso, 2004).

8. W. David Shaw, *Elegy and Paradox: Testing the Conventions* (Baltimore: Johns Hopkins University Press, 1994), 211.

9. Charles Lyell, *Principles of Geology*, Vol. II, Quoted in Eleanor Bustin Mattes, *In Memoriam: The Way of a Soul* (New York: Exposition Press, 1951), 59–60.

10. Bronislaw Szerszynski, "The End of the End of Nature: The Anthropocene and the Fate of the Human," *Oxford Literary Review* 34, no. 2 (2012): 165–82, 180. See also, Eric Gidal, *Ossianic Unconformities: Bardic Poetry in the Industrial Age* (Charlottesville: University of Virginia Press, 2015), 180–84.

11. Margaret Ronda, "Mourning and Melancholia in the Anthropocene," *Post45* 6 (2013): 1–7, 6.

12. Jahan Ramazani, *Poetry of Mourning: The Modern Elegy from Hardy to Heaney* (Chicago: University of Chicago Press, 1994), xi.

13. Timothy Morton, "The Dark Ecology of Elegy," *The Oxford Handbook of the Elegy*, ed. Karen Weisman (New York: Oxford University Press, 2010): 251–71, 255.

14. Heise, *Imagining Extinction*, 50.

15. Christophe Bonneuil and Jean-Baptiste Fressoz, *Shock of the Anthropocene: The Earth, History, and Us*, trans. David Fernbach (New York: Verso, 2015), 199.

16. Robert Pogue Harrison, *The Dominion of the Dead* (Chicago: University of Chicago Press, 2003), 142.

17. Harrison, *Dominion*, 149.

18. Eduardo Kohn, *How Forests Think: Toward an Anthropology Beyond the Human* (Berkeley: University of California Press, 2013), 212.

19. Elaine Gan, Nils Bubandt, Anna Lowenhaupt Tsing, and Heather Anne Swanson, Introduction to *Arts of Living on a Damaged Planet: Ghosts and Monsters of the Anthropocene* (Minneapolis: University of Minnesota Press, 2017), G4–G5.

20. Diana Fuss, *Dying Modern: A Meditation on Elegy* (Durham: Duke University Press, 2013), 110.

21. Devin Griffiths, *The Age of Analogy: Science and Literature Between the Darwins* (Baltimore: Johns Hopkins University Press, 2016), 155.

22. Timothy Pelatson, *Reading In Memoriam* (Princeton: Princeton University Press, 1985), 5.

23. Quoted in Erik Gray, Introduction to *In Memoriam A. H. H.*, Norton Critical Edition. (New York: Norton, 2004), xiv.

24. Gidal, *Ossianic Unconformities*, 5, 183.

25. Ibid., 182. Emphasis original.

26. Herbert Tucker, *Tennyson and the Doom of Romanticism* (Cambridge: Harvard University Press, 1988), 13.

27. Griffiths, *Age of Analogy*, 129.

28. Ibid., 130.

29. Caroline Levine, *Forms: Whole, Rhythm, Hierarchy, Network* (Princeton: Princeton University Press, 2015), 2.

30. Alex McCauley, "Alfred Russel Wallace: Media Theorist." Paper presented at the Association for the Study of Literature and Environment (ASLE) Convention. Moscow, ID. June 26, 2015.

31. Jim Endersby, *Imperial Nature: Joseph Hooker and the Practices of Victorian Science* (Chicago: University of Chicago Press, 2008), 125. "Lumpers" argued for broadly defined species and genera, whereas "splitters" argued for narrower divides, thus producing a much greater overall number of species based on local variations. It is also important to note that these images, and the species definitions they embodied, were presented in isolation, completely divorced from habitat.

32. Charles Darwin, *On the Origin of Species by Means of Natural Selection* (London: Murray, 1859), 432, 329, 312.

33. Jacques Derrida, "Signature, Event, Context," in *Margins of Philosophy*, trans. Alain Bass (Brighton: Harvester, 1982): 307–30, 328.

34. John MacNeill Miller, "Composing Decomposition: *In Memoriam* and the Ecocritical Undertaking," *Nineteenth Century Contexts* 39, no. 5 (2017): 383–98, 384.

35. Miller, "Composing Decomposition," 383.

36. Donna Haraway, *Staying with the Trouble: Making Kin in the Chthulucene* (Durham: Duke University Press, 2016), 11.

37. See George P. Landow, *Victorian Types, Victorian Shadows: Biblical Typology in Victorian Literature, Art, and Thought* (Abbington: Routledge & Kegan Paul, 1980).

38. Simon L. Lewis and Mark A. Maslin, "Defining the Anthropocene," *Nature* 519 (March 12, 2015): 171–80, 174.

39. Charles Lyell, *Principles of Geology* (1830–33), ed. James Secord (New York: Penguin, 1997), 276–77.

40. Patrick Brantlinger, *Dark Vanishings: Discourse on the Extinction of Primitive Races, 1800–1930* (Chicago: University of Chicago Press, 2003), 4.

41. Morton, "Dark Ecology," 255.

42. Brantlinger, *Dark Vanishings*, 29.

43. Ibid., 30.

44. Ashley Dawson, *Extinction: A Radical History* (New York: OR Books, 2016).

45. Henry M. Cowels traces the scientific study of anthropogenic extinction—and political advocacy to prevent it—to the work of Alfred Newton in the late nineteenth century. While it comes two decades after *In Memoriam*, Newton's work suggests that not all Victorians were as sanguine about extinction as is often supposed. Henry M. Cowels, "A Victorian Extinction: Alfred Newton and the Evolution of Animal Protection," *British Journal of the History of Science* 46, no. 4 (Dec. 2013): 695–714.

46. T. H. Huxley, "On the Formation of a Piece of Coal," *Discourses: Biological and Geological, Collected Essays*, Vol. 8 (New York: D. Appleton, 1897), 137–61. See discussion in Allen MacDuffie, *Victorian Literature, Energy, and the Ecological Imagination* (Cambridge: Cambridge University Press, 2015), 28–34, and Jesse Oak Taylor, *The Sky of Our Manufacture: The London Fog in British Fiction from Dickens to Woolf* (Charlottesville: University of Virginia Press, 2016), 127–29.

47. Arabella Buckley, *The Fairy-Land of Science* (Philadelphia: J. B. Lippencott, 1888), 192.

48. Jason W. Moore, *Capitalism in the Web of Life: Ecology and the Accumulation of Capital* (New York: Verso, 2015).

49. Dipesh Chakrabarty, "Postcolonial Studies and the Challenge of Climate Change," *New Literary History* 43, no. 1 (Winter 2012): 1–18, 14.

50. Ibid., 14.

51. I explore this idea most directly in "Globalize," Jeffery Jerome Cohen and Lowell Duckert, eds., *Veer Ecology: A Companion for Environmental Thinking* (Minneapolis: University of Minnesota Press, 2017): 30–43. See also, *The Sky of Our Manufacture*.

52. Stewart Brand, *Whole Earth Discipline: An Ecopragmatist Manifesto* (New York: Viking, 2009).

53. Will Steffen et al., "Stratigraphic and Earth System approaches to defining the Anthropocene," *Earth's Future* 4 (2016): 325–45.

54. Jeremy Davies, *The Birth of the Anthropocene* (Oakland: University of California Press, 2016). In a fittingly Tennysonian move, Davies's last chapter offers "an obituary for the Holocene" (145–92).

55. Fuss, *Dying Modern*, 7.

56. Jedediah Purdy, *After Nature: A Politics for the Anthropocene* (Cambridge: Harvard University Press, 2015), 5.

CHAPTER 3

Signatures of the Carboniferous
The Literary Forms of Coal

Nathan K. Hensley and Philip Steer

> All rightful honor, then, to these priceless Diamonds—whether they be black spirits or furnace-white, flame-red spirits, or ashy-grey—whether cannel coal and caking coal—cherry coal and stone coal—whether any of the forty kinds of Newcastle coal, or any of the seventy species of the great family, from the highest class of the bituminous, down to the one degree above old coke.
>
> —"THE BLACK DIAMONDS OF ENGLAND," *Household Words*, June 8, 1850

> Steam has been a spur to everything.
>
> —UNNAMED SHIP OWNER, 1844, quoted in *The Oxford History of the British Empire*, Vol. III, *The Nineteenth Century*

The Defined Excluded

In the 774 pages that make up Volume III of *The Oxford History of the British Empire, The Nineteenth Century*, coal is mentioned precisely three times.[1] These few sentences cast coal as barely a bit player in the grand opera of macrohistorical forces and microhistorical actors—generals, natives, economic trends, and trade arrangements—detailed in Oxford's authoritative account of the Empire. The anonymous ship owner cited above, who claimed that when it came to Empire, "Steam has been a spur to everything," is presented in the volume only as "overestimat[ing]" steam's effects.[2] Yet the mysterious substance that might conjure the mechanical power of steam was what Richard H. Horne, in an astonished *Household Words* essay of 1850, called the "priceless diamond" of Victorian modernity.[3] The Victorians knew very well that such jewels were anything but modern: Formed during the Carboniferous period, a warm and humid epoch 359 to 299 million years before human beings walked the Earth, the Victorians' black diamonds were the remains of ferns, leaves, and other organic materials subjected to pressure over vast expanses of prehuman time.[4] As water levels rose and fell, these biotic remains were buried before they could release their energy in decomposition, thus storing away "the power of millions of years of solar income . . . in a solar savings account of unimaginable size."[5] As the century progressed, therefore, Victorian England was increasingly "rooted in a past so distant it still could not be imagined."[6] Spurs to almost everything, these crystals of fossilized life had been endowed by geological luck with the capacity to do nothing less than (in Horne's words) "advance those sciences and industrial arts which are equally the consequence and re-acting cause of the progress of humanity."[7]

If coal has yet to find its place in official histories of British imperialism, this magical black stone nevertheless provided the motive power for the Empire's worldmaking project. Coal fueled the industry that made England a global power; underlay the most significant advances in technological and material progress in this most progressive age; and quite literally drove the expansionist policies of England's rapid aggrandizement and increasingly acquisitive militarization after 1880.[8] If, as Benjamin Morgan, among others, has recently observed, the Victorian period might usefully be redescribed as the Age of Coal, then the world-spanning configuration of the British Empire confirms that this energy form reigned over not only time but space.[9] Coal was the very engine of British global power in the nineteenth century, the indispensible fuel for the project of expropriation, reinscription, and extraction that Horne called "the progress of humanity." But how did the effects of this black diamond—enormous, ongoing, yet strangely resistant to conceptualization—become legible in cultural form?

In what has become our most canonical account of historical interpretation, Fredric Jameson updates a tradition of Marxist thinking about mediation to advocate a reading practice able to discover how cultural productions rearticulate the *"mode of production"* that generated them: Literary and aesthetic works come into focus as "formal conjuncture[s] through which the 'conjuncture' of coexisting modes of production at a given historical moment can be detected and allegorically articulated."[10] As is well known, this method of reading-as-decryption constitutes Jameson's key apparatus for imagining the relations of determination by which a literary "conjuncture" is construed to spring from and recast the material one contemporaneous with it. This political grounding or "ultimately determining instance" (for example, 32, 36) is the mode of production. Sophisticated as it is, Jameson's reworking of Marxian determination theory nonetheless follows its source code, in *Capital*, to see the mode of production only in light of labor relations: Thus do traces of feudalism, capitalism, and socialism, say, commingle unevenly in a given work, generating the impress of the present no less than a negative image of what is to come.

Given this focus on social relations, it is perhaps unsurprising that, as in *Capital* itself, neither coal nor any other energy form earns significant mention in *The Political Unconscious*. But as this essay will show, attention to energy regimes helps us appreciate that the mode of production that even our most persuasive theories of mediation view as the elemental "level" in any system of social mediation—its ultimately determining instance, or what Marx calls the "absolutely objective conditions" of an "economy"—is itself subtended by another "level," an energy regime with respect to which the political itself is, as it were, superstructural.[11] Raymond Williams and Louis Althusser, among others, have helped trouble this language of levels and planes, bases and superstructures, and have shown how the relations among seemingly separate domains of historical experience are far from simple, stratified, or easily hierarchizable: They are, in Althusser's term, "overdetermined."[12] Still, it remains the case that to raise the problem of energy's relation to "production" is to reanimate the oldest problems in materialist criticism but locate them, as it were, deeper; and we might follow Tobias Menely, Jason Moore, and others in seeking to understand how the canonical problem of determination becomes unspooled and reorganized with attention to systems of energy and the yet more elaborate models of historical causality they challenge us to imagine.[13] These dilemmas become further complicated when we ask how a system of energy storage, transport, and conversion that is structuring and omnipresent, even if unevenly distributed, and therefore all but impossible to conceptualize as such from within, becomes visible in cultural productions seemingly unable or unwilling to engage this energy system, *as* a system, directly. After all, as Jameson argues in *Political Unconscious* with respect to the relationship between the text and the "social ground" from which it emerges, "the social contradiction addressed and 'resolved' by the formal prestidigitation of narrative must . . . remain an absent cause, which cannot be directly or immediately conceptualized by the text" (75, 82). Jameson's later analysis of life under global capitalism explains how the "structural coordinates" of daily experience are "no longer accessible to immediate lived experience and are often not even conceptualizable for most people."[14] If this is true, then by what indirect means did the infrastructure of coal-life emerge into form? And if coal was and remains the disavowed force behind Victorian modernity, its spur to everything, by what methods might we discover its signature?

This chapter revises existing accounts of Victorian mediation by locating what is arguably the signal cultural form of the nineteenth century—the novel—within the global energy system that increasingly made it possible. While we engage political and economic theory, we here leave aside epic poetry, oil painting, journalism, photography, theater, and dance—along with myriad other cultural forms whose shapes, logics, and formal designs would have been decisively shaped, in some way or another, by the effects of coal. (Print journalism is just one obvious place where coal becomes legible as form, since the literal shape of the journalistic article changed based on advances in steam-driven printing presses.) Our aim in this deliberately constrained experiment in reading for coal is to offer a test case in adducing how the practices and infrastructures of fossil combustion became legible as literary effect.

Writing of the oil-based economy of the twentieth and twenty-first centuries, Ross Barrett and Daniel Worden have described the "curious valence" of oil in the "cultural imagination," whereby it is "not invisible to us as much as it is contained—in our cars' gas tanks, in pipelines, in shale, in tar sands, in distant extraction sites."[15] Coal is likewise obliquely omnipresent in Victorian literature. Dickens's account of the construction of the London-Birmingham railway in *Dombey and Son* (1848)—where, famously, the railway, "from the very core of all this dire disorder, trailed smoothly away, upon its mighty course of civilisation and improvement"[16]—is memorable in part because of its anomalous interest in the social, spatial, and economic "earthquake" produced by steam. More common are cultural forms that depict railway journeys as an ordinary part of their narrative lifeworlds; more common still are those that, while alluding to steam-powered travel or the products of steam-driven manufacture, regard these aspects of narrative infrastructure as entirely beneath the interests of story: They melt the socio-environmental processes of energy extraction, storage, combustion, and conversion, almost reflexively, into the category of the everyday.[17] In this sense do steam and the coal that fired it become recognizable as what Althusser refers to as a "condition of possibility" within a historical structure, one that, precisely *because* it undergirds all facets of experience within what he calls a historical field or "problematic," is inapprehensible from within it: In Jameson's words quoted previously, which channel Althusser, it is a "truth" that is "no longer accessible to lived experience."[18] Approached this way, coal is what Althusser calls a "defined excluded," something "*excluded* from the field of visibility and defined as excluded by the existence and particular structure of the problematic."[19] A society that depended entirely on coal could barely, precisely because of that dependence, become conscious of coal at all.

How did this darkness become visible? How did the unrepresentable find shape? The pages that follow propose one way of answering those questions, by attempting what we term a hermeneutics of coal. E. A. Wrigley has argued that the Victorian era saw a coal-driven transition from an "advanced organic economy" to a mineral-based "energy economy."[20] In this historical shift, economic growth became decoupled from the limits of agricultural production for the first time in history. Given the unmooring of productive power under the coal regime, we argue two related points about coal form. First, coal plays a structuring role in texts that consider how bounded or localized systems of belonging—economies, nations—might be transgressed, opened up, or otherwise superseded. The spectacular energy potential of fossil carbon, in other words, was the enabling condition for an increasingly global imaginary. Second, we suggest that the scope of those carboniferous literary effects becomes fully apprehensible only when we constellate texts from across the full expanse of the era's carbon-fueled economy, in what are usually conceived as discrete categories of genre and geography.

Chasing coal's signature, this chapter telescopes from the canonical scenes of Victorian extraction and combustion that criticism has long filed under the heading "industrial"—the metropolises of England's northern counties—to colonial peripheries rarely included in extant stories of British coal. We begin with a diptych of coal-haunted novels by Elizabeth Gaskell, and set the archetypal industrial romance, *North and South* (1855), against the sketchy and all-but-plotless *Cranford* (1853); turn to J. R. Seeley's romance-inflected manifesto for an Empire-wide British polity, *The Expansion of England* (1883); and conclude with Joseph Conrad's auto-demolishing analysis of extractive capitalism at the Pax Victoriana's violet hour, *Nostromo* (1904).

"Friends in this big smoky place"

In *North and South*, we find coal's signature not only on its familiar scenes of urban squalor and industrial exploitation, but in the novel's (impossible and unsatisfied) desire to find narrative closure in the organic form of the nation. Fissured by railways, Irish migrant labor, volatile American supplies of cotton, and the fluctuations of global credit, the novel stages the nation's new coal-powered networks as structurally unimaginable even as they are materially unavoidable: "By the 1840s, coal was providing energy that in timber would have required forests covering twice the country's area, double that amount by the 1860s, and double again by the 1880s."[21] Viewed from the perspective of Victorian energy regimes, the novel's structuring opposition—between agricultural South and industrial North—comes into focus as a confrontation between (1) the traditional organic economy, in a static state deriving from the need to "live within limits set by their ability to capture some fraction of a [solar] flow whose size varies very little from year to year," and (2) a new, coal-fired economy driven by "*stocks* of energy rather than [built] upon organic energy *flows*."[22] Manchester, fictionalized by Gaskell as Milton-Northern, was ground zero for this transformation: Fueled by the vast coalfields in neighboring Lancashire, more than 500 chimneys choked the city by the 1840s, the smoke a byproduct of booming cotton production; the city's population had more than quadrupled in half a century to more than 300,000 by 1851; in their homes, those residents were burning an estimated two million tons of coal annually, or approximately five tons per capita.[23]

In the earlier *Cranford* (1853), Gaskell had taken the rapidly altering social and geographical provincial landscapes of her carbon economy as the occasion to unravel the architecture of the novel: this book, first published in a run of essayistic entries in *Household Words*, became as a novel a series of plotless sketches, its form a vectorless equilibrium punctuated by bank failures (which ruin Matty), allusions to the imperial deathworld of India (where Peter falls ill and expires), and the *deus-ex-machina* of the "nasty cruel railroads" (which run over Mr. Brown, who prefers *Pickwick Papers* to Dr. Johnson, as he is distractedly "a-reading some new book as he was deep in").[24] The only references to coal in this inward-looking text direct us to the domestic hearth. Still, in its self-reflexive nods to popular fiction—which Dickens famously altered in the serial version, removing the *Pickwick* reference—and a wider world beyond its pages, Gaskell labors to connect her own fictional practice to the railway economy at its full global scale, going further to mark this steam-driven economy and the literature proper to it as tracking toward death, anomie, ruin. The book's seemingly isolationist naiveté is undercut by a globalizing irony because "such simplicity might be very well in Cranford, but would never do in the world."[25] By contrast to this enigmatic modernity tale, Gaskell's archetypal industrial novel, *North and South*, unfolds in an affirmative mode, figuring the new carbon-based speculative and imperial economy through Margaret Hale's domesticating encounter with Milton-Northern. Here, Gaskell maps the intersections among population, urban geography, and economics in ways no less detailed than in *Cranford*; but *North and South*'s setting—in the metropole rather than the provinces—means that the residue of the force binding all these factors together, coal, hangs over the novel's world:

> For several miles before they reached Milton, they saw a deep lead-coloured cloud hanging over the horizon in the direction in which it lay. ... Nearer to the town, the air had a faint taste and smell of smoke; perhaps, after all more a loss of the fragrance of grass and herbage than any positive taste or smell. Here and there a great oblong many-windowed factory stood up, like a hen among her chickens, puffing out black "unparliamentary" smoke, and sufficiently accounting for the cloud which Margaret had taken to foretell rain.[26]

Margaret's first impression shows us that Milton-Northern is, from the outset, imagined as an ecosystem of the carbon economy. In the absence of nature, the novel presents the environmental question of air quality as inseparable from the spatial reorganization of the city's residential areas and the prominence and power of industrial production. As Barbara Freese observes, workers' lives in industrial cities such as Manchester or Milton-Northern were "constructed, animated, illuminated, colored, scented, flavored, and generally saturated by coal and the fruits of its combustion."[27] In mid-century urban centers such as the one Gaskell documents, then, coal was both phenomenological horizon (because everything one could experience was "saturated" with it) and total institution (because there was no escape from its effects). Yet the totalizing fact of coal-life registers only slightly in Gaskell's novel; once Margaret is immersed in this milieu, coal is barely mentioned, and references to the city's smoky air fade to insignificance. As direct notation falls away, the novel's sensitivity to processes of coal-fired social reorganization reconstitutes itself in the language of energy, strength, and power that pervades its account of the city, and especially Margaret's consciousness of the Byronic factory owner, John Thornton.

In ways perfectly foreign to the queer and sexless *Cranford*, *North and South* uses Margaret's erotic attraction to Thornton to imagine the rising industrialist class as a potentially fecund marriage between brute masculine productivity and domestic manners, fetchingly female. But this union also enables what Thornton describes as the "imagination of power" by personifying, and thus domesticating, the effects of the combustion of coal (81). The characterization of the brooding, "Teutonic" mill owner enables steam technology to be masculinized and eroticized as the conqueror of a passive "inanimate nature": "rather rampant in its display," the new form of power now able to be commanded "seemed to defy the old limits of possibility" (162). Margaret's susceptibility to Thornton and his rampant machinery—that is, the allegorizing of the carbon economy through the love plot—effectively naturalizes the coal-fired economy, presenting it as something that merely requires a more respectful treatment of its laborers to be accommodated by the existing organic imagined community. Whereas Margaret had previously been concerned that Thornton's home was unhealthily close to his place of business, "blackened, to be sure, by the smoke, but with paint, windows, and steps kept scrupulously clean," now industrial harmony is found in the transformation of the factory into a domestic space, as he constructs a dining room for his employees and they more than return the favor by voluntarily working overtime (111). Yet the irony persists that the novel's ability to imagine a unified nation is predicated upon Margaret's use of the same coal-powered technology that is pulling it apart and reshaping it. Although it is "[r]ailroad time" that first "inexorably wrenched them away from lovely, beloved Helstone," it is also the railway's ability to bridge the North and the South that nevertheless allows Margaret to comprehend them within a single frame at all (57).

The novel thus labors to domesticate the very forces whose catastrophic unleashing it documents. These forces are global in nature, and over the course of the narrative, the novel proves unable to contain the far-flung threats to national stability that have been brought about by the coal-powered annihilation of distance. Andreas Malm argues for the necessity of understanding the intersection of "thermodynamic and social power" in the use of fossil-fuel energy for, "by definition," they are "a materialisation of social relations."[28] As *North and South* documents, coal allowed mill owners to transcend Britain's borders in search of profits, whether by threatening to relocate their operations if labor costs rise further or by importing migrant labor from across the Irish Sea to undermine the conditions afforded to local laborers. Yet by the novel's conclusion, its perspective on the global marketplace is itself transformed, as what had at first only been figured as a source of raw materials and consumer demand—the global market itself—is ultimately revealed to be so powerful and destabilizing that any national rapprochement between the masters and men ultimately appears to be only a temporary solution at best. Despite Thornton and his employees' finding a way "to look upon each other with far more charity and sympathy," it takes only a few pages for the fluctuations of a global market to leave Thornton, like Matty in *Cranford*, ruined: forced to "give up the business in which he had been so long engaged with so much honour and success" (410, 415). In this way has coal split apart the very literary and social forms (the novel, the nation) Gaskell marshals to contain its energies.

In *The Industrial Reformation of English Fiction*, Catherine Gallagher finds factory literature from the industrial decades—narratives such as *Hard Times* (1854), *Michael Armstrong* (1840), *The White Slaves of England* (1853), and *North and South* itself—to be organized around what she calls "tropes of reconciliation": more or less elaborate formal and ideological solutions whereby the public (male) world of wage labor and market capitalism is made by means of plot to harmonize with the private domestic (female) world of the family that is its natural antagonist. But if the nineteenth-century novel is defined, as Gallagher shows, by the "structural tension between impulses to associate and to disassociate public and private realms of experience," reading for coal's signature shows that the form is called upon to manage yet more profound structural tensions than these.[29] The work of the industrial novel, we argue here, is to "manage" the new energy regime that made its very existence possible. In the hands of Gaskell and other writers of industrial romance, in other words, the technology of plot becomes the means by which the horizonless potentialities of coal might provisionally or aspirationally be bound, contained, and made thinkable within a national paradigm defined, now, by the marriage plot and its implicit corollaries, heterosexual domesticity and reproductive futurity. *Cranford*'s queer plot, refusing each of these solutions in turn, ends its seemingly ateleological meander with Matty, unmarried and nonreproductive, scraping together a locally scaled business indifferent to the utilitarian calculus of profit. *North and South* builds to a more conventionally satisfying conclusion, with Margaret providing Thornton with a welcome infusion of capital that allows him to return to his role as mill owner, their financial and erotic plots ultimately sealed as one. Yet this heavily freighted marriage plot, cancelling social antagonism and ensuring Thornton's continued ability to extract profit from the system of carbon-fired exploitation he oversees, can do nothing to address the destabilizing international economic shifts, always tending toward ruin, coal will eventually bring home.

Unparalleled Expansion

North and South deploys romance to offer its recuperative response to the social and economic forces unleashed by the steam-driven economy: the erotics of the marriage plot wrest from this chaos a fantasy of the racially pure, harmonious, and future-oriented nation. In subsequent decades, a similar reflex toward containment underpinned the most forthright attempts to imagine the political effects of coal technology at an imperial scale. The theoretical vanguard of efforts to manage the endlessness and unfixability of this new form of capital accumulation was located within the movement to create a global British polity: as J. R. Seeley put it in *The Expansion of England* (1883), "Science has given to the political organism a new circulation, which is steam, and a new nervous system, which is electricity.... They make it in the first place possible actually to realise the old utopia of a Greater Britain, and at the same time they make it almost necessary to do so."[30] Seeley's best-selling history, self-avowedly "haunted by the idea of development, of progress" (3), argues that English history has, for the last few centuries, primarily occurred offshore, and that England's most distinctive political innovation during that time has not been Reform or Liberalism but "a peculiarly English movement... [of] unparalleled expansion" (308). In this account, settler colonialism constitutes a natural extension of the English state, unified by race and language and an apparent absence of natives, cleansed too of the despotic traits associated with ruling India. Yet the smooth surface of Seeley's tendentious nationalist tale is sporadically ruptured by recognition that the expansive tendencies he describes are the centripetal force intrinsic to the coal economy. At such moments of splintering narrative coherence, steam transportation emerges as causal rather than merely catalytic in the process of invasion. Empire is impossible without coal, in other words, and we are only now waking to their joined splendors: "Perhaps we are hardly alive," Seeley writes, "to the vast results which are flowing in politics from modern mechanism" (299). The settler Empire thus functions in *Expansion of England* less as English national destiny than as a temporary "spatial fix," in David Harvey's term, for the political and cultural contradictions of extractive capitalism. Settler colonialism offers a necessary alibi to Empire, a comforting myth of limitless resources untainted by violence. Accordingly, Seeley's encomia to steam directly abut his accounts of settler polities as natural, familial, "normal": "[W]e see a natural growth, a mere normal extension of the English race into other lands, which for the most part were so thinly peopled that our settlers took possession of them without conquest. If there is nothing highly glorious in such an expansion, there is at the same time nothing forced or unnatural about it" (296). Here Gaskell's romance of the organic national community is writ large, as the antagonisms and crises of the international fossil imperium are tied up in the bow of "natural" domesticity. "The tie that holds together the parts of a nation-state," we are told, "is not composed of considerations of profit and loss, but ... analogous to the family bond," an expanded England proffered in an attempt (unevenly) to synthesize and contain the global violence of the carbon economy (63). Seeley places Britain at the geographic center of the discussion and posits the periphery of Empire as empty, available space, an unconsidered site of extraction and promise whose violation, forced and rapacious, is inconceivable from within the terms of his argument. Yet as the century charged toward its twilight hour, the spatial and temporal limitations imposed on British power by its dominant source of energy would become increasingly apparent. And the dissolution and anomie lurking within the carbon economy, at once the precondition and end result of the very structure of expansion Seeley advocates, would soon be impossible to ignore.

Treasure from the Earth

Concerned directly with national myth and global expansion under extractive capitalism, Conrad's *Nostromo* aims to radicalize rather than resolve the impossible dilemmas of coal form. The novel is populated with a series of characters who think along lines laid out for us by Gaskell and Seeley. On the one hand are the would-be nation-builders: European characters like Charles Gould and Martin Decoud whose cause, in their adopted homeland, is to carve stability and civil society out of a war-ravaged and serially revolutionary extraction zone. On the other hand are the theorists of endless expansion, notably represented by the American financier Holroyd, who remotely funds the operation of the San Tomé mine and is thus the meta-sovereign behind even Gould, that (oft repeated) "Rey de Sulaco." "We [Americans] shall be giving the word for everything," explains the real king, Holroyd, in a famous line: "industry, trade, law, journalism, art, politics, and religion, from Cape Horn clear over to Smith's Sound, and beyond, too, if anything worth taking hold of turns up at the North Pole."[31] This financier's promise of an American-led universalism updates not just Seeley's Greater Britain but Cecil Rhodes's often quoted desire to annex the stars; it also signals Conrad's interest in parsing the inter-imperial or transitional moment his novel documents, at the dawn of the American Century and the waning days of British global hegemony. In the novel, this shifting geopolitical situation takes shape as plot, as the residually aristocratic Charles Gould, from England, must partner with the "Steel and Silver King" (173) in San Francisco, who actually pulls the strings on Gould's extractive enterprise.

The pawn at stake in this macropolitical struggle is the seaboard state Conrad names Costaguana. The name alludes to its richness in that early-to-be-exploited biogenic resource—guano—and also signals that since its earliest days, this place has yielded its natural resources for the benefit of those elsewhere. The primary form of geophysical treasure in the novel is, of course (as one section title calls it), "The Silver of the Mine." Yet a host of other commodities—gold, guano, copper, and even ox hides—are stripped from the hillsides and converted into value. It falls to Mrs. Gould to notice, just glancingly, the catastrophe on which such investment opportunities are predicated. She "had seen it all from the beginning; the clearing of the wilderness, the making of the road, the cutting of new paths up the cliff face of San Tomé" (80). And where a waterfall had once been was now, after this development, "only the memory of the waterfall": "The tree-ferns that had luxuriated in its spray had dried around the dried-up pool, and the high ravine was only a big trench half filled up with the refuse of excavations and tailings" (79). Charles Gould, his own name echoing the metal Spanish galleons had once stripped from the territory, refers with uncertain tone to his own work as "the tearing of the raw material of treasure from the earth" (46)—albeit as he visits Italy, and tours a marble quarry. The detail confirms that the novel's critique of extractive capitalism is comprehensive: "Tomé," as Nasser Mufti observes, means "to take."[32] But Gould's mission, like Thornton's in *North and South*, is improvement. Before Gould takes over the San Tomé mine, it had fallen into disrepair:

> Worked in the early days mostly by means of lashes on the backs of slaves, its yield had been paid for in its own weight of human bones. Whole tribes of Indians had perished in the exploitation; and then the mine was abandoned, since with this primitive method it had ceased to make a profitable return, no matter how many corpses were thrown into its maw. (40)

Conrad's layered prose ensures that "the exploitation" refers syntactically to the financial kind. But the word echoes in the (physical) exploitation of native bodies under slavery. All of it adds up to what the novel calls "the sordid process of extracting metal from under the ground" (41).

Nostromo's most unmistakable lesson may be, as Marx had long before noted, that all value begins in blood. Yet the details showing us the spectacular violence of primitive accumulation also confirm that Charles Gould's competitive advantage—what enables him to turn a profit from this formerly unprofitable, slave-worked mine—is yet another extracted commodity, this one all but unmentioned in the novel. It is, after all, Gould's steam-driven railroad, "dug [from] the earth [and] blasted [from] the rocks" (28), that the novel specifies is the first step in restoring the San

Tomé mine to profitability, meaning that the energy regime of coal stands as the final, if curiously spectral, material interest driving this dependency-state development narrative. Conrad gives this determining agency a ghostly presence, a semi-visibility that comes into focus most, perhaps, with the novel's obsessive attention to "steam": The word or its variants, such as "steamer," appear 70 times, describing steam-driven mail boats, U.S. warships ironically named after native tribes ("Powhattan"), or the railroad crucially linking San Tomé to Sulaco. The primary usages are nautical, and like so many of Conrad's other novels—*The Nigger of the "Narcissus"* (1897), *Heart of Darkness* (1899), and *Typhoon* (1902) in particular—*Nostromo* foregrounds the historical transition between sail and steam and uses this moralized dichotomy of maritime energy (sail good, steam bad) to critique the noisy modernity of coal-fired travel. In *Nostromo*, the two regimes come into brutal contact, literally crashing together when, in one of the novel's key episodes, the silent and sail-driven lighter commanded by Nostromo is smashed by a chugging steamer helmed by General Sotillo (210).

The crack-up condenses into allegory the historical switch whereby an organic, romanticized imperial mode, typified by silence and sail and nature, is overtaken and indeed smashed to bits by the cacophonous modernity of a coal-fired steamer. Conrad's ideological investment in residual energy forms again becomes legible when, in *Nostromo*'s opening pages, we learn that until the dawn of steam power, Sulaco had enjoyed an "inviolable sanctuary from the temptations of a trading world" (5), as sailing vessels and Spanish galleons were kept out of the harbor by the "atmospheric conditions" of "variable airs" (9). But these winds "could not baffle the steam power of [the Oceanic Steam Navigation Company (O.S.N.)'s] excellent fleet" (9). With the Company's ships named for Greek gods, steam is, from *Nostromo*'s outset, construed as a hypersexualized and perversely divine force, able to pierce once-natural boundaries and ravish formerly pristine landscapes. The repeated mantra of this "Tale of the Seaboard" is that Sulaco is the "treasure house of the world" (344, 347, 351), but the treasure house is unlocked with coal.

The brutality of the novel's steam-driven progress is manifest, and Conrad's irony obvious. But for Gould and the European characters like Decoud and Viola who believe in the possibility of progress, the hope is that, as Gould puts it, "a better justice will come afterwards" (63). Gould's lines about "law, good faith, order, security" requiring "material interests" to "get a firm footing" (63) could have been ripped from the pages of Seeley or any other bourgeois theorist of English imperialism. (Signaling this, Conrad has Gould give this speech in "his English get-up" [63].) It is axiomatic, in

Gould's civilizational narrative of capital, that economic exploitation must precede the establishment of legitimate government. The silver of the mine, we are told, will have a "justificative conception" (80). Until that stability arrives, however, Costaguana appears as an endless series of meaningless wars and fruitless revolutions, its would-be saviors dying in squalid shootouts (like General Montero), suicides (like Decoud), or absurd misunderstandings in the night (like Nostromo). Everyone in Costaguana was being killed, so Mrs. Gould hears, in "battles of senseless civil wars, barbarously executed in ferocious proscriptions, as though the government of the country had been a struggle of lust between bands of absurd devils let loose upon the land with sabres and uniforms and grandiloquent phrases" (66). Against these cycles of political violence—constitution, dissolution, and reconstitution, all in a sequence—stability is impossible: No founding myth, be it a marriage plot (as in *North and South*) or nationalist ideology (as in *Expansion of England*), can still the permanent motion of extractive international capitalism. "The arc of Costaguana's history," Mufti writes, "is all crisis with no moment of arrival."[33]

Conrad's achievement is thus to radicalize the non-progressive vision Gaskell offers in *Cranford*. As in that almost formlessly episodic sequence of set-pieces, Conrad offers the antidevelopmental or properly cyclical historical model endemic to modernity's sacrifice zones as a finally narrative or formal dilemma: *Nostromo* folds the endlessness of fossil-capitalism's structure into its own narrative presentation, crystallizing coal as form. *Nostromo*'s spiraling temporal structure, that is, matches coal's fixity-splitting tendencies to the novel's narrative procedure, defeating dreams of progress and equilibrium at narrative and historical levels alike. *Nostromo*'s formal difficulties are famous. Its endless series of revolutionary failures spins and spins, as cycles of anti-teleological historical motion radiate outward and repeat, the novel compounding prolepses and analepses in nested sets of flashbacks and flash-forwards that are, for many readers, almost impossible to parse. These acrobatic temporal effects fuse time into odd and nonlinear configurations, and have driven critics to cite "[t]he novel's much discussed and often confusing time shifts," and conclude that "there is no other Conrad work ... that flaunts problems of temporal displacement and deferral, and challenges assimilation to any specific moment 'in' time, the way this one does."[34] One contemporary observed that "it is often difficult to say when or where we are" in the plot, and the book's modern editors admit that it "cannot be read unless one has read it before."[35] ("The novel ends, in a sense, where it began."[36])

In its very form, then, this novel of endless revolution confirms Hannah Arendt's sense, in *On Revolution* (1963), that political overturning presents special difficulties for narrative structures dependent on closure. (Recall that *Cranford*'s final chapter, delicately ironic, is "Peace to Cranford.") For Arendt, "the modern concept of revolution" is "inextricably bound up with the notion that the course of history suddenly begins anew, that an entirely new story, a story never known or told before, is about to unfold."[37] Like *Cranford*, *Nostromo* translates the serialized, ruinous, and open-ended logic of carbon modernity into plot, and, like *Cranford*, discloses how a narrative infrastructure might both derive from and be implicated in the effects of the coal economy it seemingly only documents. In *Nostromo*, such moments of new beginning as Arendt describes are repeated serially, endlessly, so that the very novelty of new beginning itself becomes a perverse or traumatic repetition. For *Nostromo* no less than for *Cranford*, this futureless stasis—"sterile," as Edward Said calls it—is allegorized through an environment in which heteronormative sex is thwarted, avoided, or canceled, reproductive futurity sterilized into a parade of set-pieces.[38] Conrad's novel goes further than Gaskell's to suggest that this simultaneously political, sexual, and formal predicament issues from the very logic of extractive accumulation on which the political situation it documents is based: In extractive economies, no fix is possible, no stillness in sight, no viable future imaginable.

Like its sexless and often remarked hypermasculinity, then, *Nostromo*'s immense chronological difficulty derives from and in turn explains the unfixable surplus generated in extractive imperialism. Superadded to all this, the novel's structure of irony means that even when its own conclusion is heroically announced—as it is by Captain Mitchell, in the ostentatiously confident speech of national pride at the end of the novel—this apparent resolution into stability is instantly undercut, and the book's looping sequence, common to permanent political revolution no less than to psychic trauma, never does come to rest. The very grammar of Mitchell's speech, rendered in the habitual past (he "*would* lead some privileged passenger" [341], "*would* keep on talking" [341], "*would* talk" [342], and "*would* say" [343]) exposes the recycled and even endless repetition of his performance, even as the novel seems to mark the nationalist speech as a hyper-particularized individual instance, happening just once, as he (for example) "hold[s] over his head a white umbrella with a green lining" (343). When the novel refers twice to "the cycle" of Mitchell's own story (345, 350), it hints that Mitchell's reiterated or nearly reiterated discourse is, like everything else in the novel, a repeat performance. Like Gaskell's ironized call for "Peace to Cranford," Mitchell's paean to the stable achievement of "The Occidental Republic" contains the specter of that Republic's dissolution. What this tells us is that cycles of exploitation in the sacrifice zones of Conrad's carbon modernity will never resolve into stability, however much Gould or Mitchell might dream (with *North and South*) that he has "closed the cycle" (351).

Opening the Cycle

If the most overt task of this essay has been to return to the Victorian novel with an awareness born of our own carbon-saturated atmosphere, a corollary effort has been to use the defined excluded of coal infrastructure to unsettle or reorient our own critical categories, to open a "beyond" to even our most sensitive methods for dialectical reading. Yet the point is that such vertiginous, second-order thinking is precisely what Jameson himself announces as criticism's most important task. "[D]ialectical thinking," he explains in a famous sentence, "is a thought to the second power, a thought about thinking itself, in which the mind must deal with its own thought process just as much as the material it works on, in which both the particular content involved and the style of thinking suited to it must be held together in the mind at the same time."[39] We are now struck by the manner in which coal seems to have also infused, invisibly yet pervasively, our critical heritage, shaping not just "literary form" but the form of thought itself, even our own, now, at the very moment we write. In a fascinating instance at the heart of the first long chapter in *The Political Unconscious*, "On Interpretation: Literature as a Socially Symbolic Act," Jameson cites Max Weber on the topic of bureaucratic society and its "iron cage," and (using Weber's ventriloquized language) inadvertently alludes to the sphere of carboniferous energy conversion we've tracked here, otherwise absent from this most sophisticated account of social mediation. The Puritan order, Jameson lets Weber tell us, "is now bound to the technical and economic conditions of machine production which today determine the lives of all the individuals who are born into this mechanism, not only those directly concerned with economic acquisition, with irresistible force. Perhaps it will so determine them until the last ton of fossilized coal is burnt" (90n). Cited, consigned to a note, and rendered in another's language, this reference to coal-based "determin[ation]" could be viewed as a Derridean supplement to our most canonical account of historical interpretation, its defined excluded. The absent cause of burned carbon reappears yet more strikingly in "Modernism and Imperialism." There, Jameson argues that the most characteristic formal effects of modernist literature—an impulse toward mapping, spatial derangements, and a protocinematic crosscutting or montage among them—derive fundamentally from the imperial predicament, because this globalized material scene introduces a "spatial disjunction" by which metropolitan subjects become unable cognitively to grasp their world system in its totality. Constellating Conrad with Seeley and Gaskell, across genres and standard periods, has introduced us to a way of seeing how an apparently "modernist" form might not be modernist at all. That is because it derives not just from the "imperialist dynamics of capitalism proper," but from the coal-age energy sources without which those dynamics would never have been possible.[40] This new attention to energy systems might, in turn, help us appreciate why, in one of Jameson's signal exhibits of modernist form, from *Howard's End*, Mrs. Munt speeds through the English landscape on a train, frantically raising and lowering her window to avoid inhaling the residue of the fuel even Jameson cannot yet name.[41]

Notes

1. In his comprehensive evaluation of the role played by coaling stations in the nineteenth-century Empire, Steven Gray notes that there are more references to coconuts and coffee than to coal in the volume. "Black Diamonds: Coal, the Royal Navy, and British Imperial Coaling Stations, circa 1870–1914" (PhD thesis, University of Warwick, 2014), 17.

2. Robert Kubicek, "British Expansion, Empire, and Technological Change," in *The Nineteenth Century*, ed. Andrew Porter and Alaine M. Low, *The Oxford History of the British Empire* (Oxford: Oxford University Press, 1999), 249.

3. Richard H. Horne, "The Black Diamonds of England," *Household Words*, June 8, 1850, 248.

4. Coal's origin in plant decomposition came to be accepted in Britain between the end of the eighteenth century and the publication of Charles Lyell's *Principles of Geology* in the 1830s. Andrew C. Scott, "The Legacy of Charles Lyell: Advances in Our Knowledge of Coal and Coal-Bearing Strata," in *Lyell: The Past is the Key to the Present*, ed. Derek J. Blundell and Andrew C. Scott (London: Geological Society, 1998), 244.

5. Barbara Freese, *Coal: A Human History* (London: Heinemann, 2003), 6.

6. Freese, *Coal*, 69.

7. Horne, "Black Diamonds," 248.

8. The age of equipoise was also the age of sail; as the navy transitioned to a primarily steam-powered fleet in the latter decades of the century, expansion

became the rule. Hobsbawm notes that the 22,000 steamships in the world by 1882 were more than surpassed in tonnage by sailing ships; yet this balance would change "immediately and dramatically" in favor of steam in the 1880s. As the final section of this chapter notes, this transition is among the central concerns in Conrad's oeuvre. E. J. Hobsbawm, *The Age of Empire, 1875–1914* (New York: Vintage, 1989), 28.

9. Benjamin Morgan, "*Fin du Globe*: On Decadent Planets," *Victorian Studies* 58, no. 4 (2016): 610.

10. Fredric Jameson, *The Political Unconscious: Narrative as a Socially Symbolic Act* (Ithaca: Cornell University Press, 1981), 89, 99. Hereafter cited parenthetically in text.

11. Marx quoted in Raymond Williams, *Marxism and Literature* (New York: Oxford University Press, 1977), 85. As Patricia Yaeger writes, "thinking about literature through the lens of energy, especially the fuel basis of economies, means getting serious about modes of production as a force field for culture." "Editor's Column: Literature in the Ages of Wood, Tallow, Coal, Whale Oil, Gasoline, Atomic Power, and Other Energy Sources," *PMLA* 126, no. 2 (2011): 308.

12. See for example, Raymond Williams, "Determination," in *Marxism and Literature* (Oxford: Oxford University Press, 1977). Louis Althusser introduces the concept of "overdetermination," a figure borrowed from Freud, to name the intricate and ultimately unknowably tangled set of causal factors affecting any historical configuration or event. *Reading Capital*, trans. Etienne Balibar and Ben Robert Brewster (London: Verso, 1979), 188. Williams suggests that the "levels" of social mediation cannot even properly be construed as "levels" at all because the allegedly immaterial or superstructural forms of culture are themselves material, while the "concrete" sphere that Jameson names the political is importantly shaped—we might say determined—by cultural or "superstructural" factors of all kinds. The Victorian novel, generated amidst a rapidly expanding industry for culture and shaped by advances in material production like printing presses, arguably brings this interpenetration of "culture" and "materialism" most clearly to the fore.

13. Tobias Menely, "Anthropocene Air," *Minnesota Review* 2014, no. 83 (2014): 93–101; Jason W. Moore, *Capitalism in the Web of Life: Ecology and the Accumulation of Capital* (London: Verso, 2015).

14. Fredric Jameson, *Postmodernism; Or the Cultural Logic of Late Capitalism* (Durham: Duke University Press, 1991), 411.

15. Ross Barrett and Daniel Worden, "Introduction," in *Oil Culture*, ed. Ross Barrett and Daniel Worden (Minneapolis: University of Minnesota Press, 2014), xvii.

16. Charles Dickens, *Dombey and Son*, ed. Peter Fairclough (London: Penguin, 1970), 121.

17. Nicholas Daly points out that "by the 1860s railway travel had been almost completely assimilated into everyday life." Nicholas Daly, "Railway Novels: Sensation Fiction and the Modernization of the Senses," *ELH* 66, no. 2 (1999): 471. In Trollope's *Phineas Finn*, which traces its protagonist's frequent movements between London, Ireland, and Scotland, Phineas's steam train and steamship journeys are barely alluded to, while those modes of transport are treated as synonymous with the unremarkable: Lord Chiltern's proposal to Violet Effingham is doomed because he "asked her to be his wife as a man asks for a railway-ticket or a pair of gloves, which he buys with a price." Anthony Trollope, *Phineas Finn* (Oxford: Oxford University Press, 1973), 314.

18. Althusser, *Reading Capital*, 25. Jameson refers to imperial globalization but not the energy infrastructure that fueled it. *Postmodernism; Or, the Cultural Logic of Late Capitalism* (Durham: Duke University Press, 1991), 411. Our contention is that coal infrastructure is, like Jameson's globalization, an all-but-unconceptualizable condition for phenomenological life.

19. Althusser, *Reading Capital*, 26, emphasis original.

20. E. A. Wrigley, *Continuity, Chance and Change: The Character of the Industrial Revolution in England* (Cambridge: Cambridge University Press, 1988), 68–97.

21. Timothy Mitchell, "Hydrocarbon Utopia," in *Utopia/Dystopia: Conditions of Historical Possibility*, ed. Michael D. Gordin, Gyan Prakash, and Helen Tilley (Princeton: Princeton University Press, 2010), 120.

22. Wrigley, *Continuity, Chance and Change*, 51, 55.

23. Stephen Mosley, *The Chimney of the World: A History of Smoke Pollution in Victorian and Edwardian Manchester* (Abingdon, Oxon: Routledge, 2008), 14–19.

24. Elizabeth Gaskell, *Cranford* (New York: Penguin University Press, 2005), 22, 23.

25. Ibid., 170.

26. Elizabeth Gaskell, *North and South*, ed. Patricia Ingham (London: Penguin, 1995), 60. Hereafter cited parenthetically in text.

27. Freese, *Coal*, 73.

28. Andreas Malm, *Fossil Capital: The Rise of Steam-Power and the Roots of Global Warming* (London: Verso, 2016), 18, 19.

29. Catherine Gallagher, *The Industrial Reformation of English Fiction: Social Discourse and Narrative Form, 1832–1867* (Chicago: University of Chicago Press, 1985), 149.

30. J. R. Seeley, *The Expansion of England: Two Courses of Lectures* (London: Macmilllan, 1883), 74. Hereafter cited parenthetically in text.

31. Joseph Conrad, *Nostromo: A Tale of the Seaboard*, ed. Jacques A. Berthoud and Mara Kalnins (New York: Oxford University Press, 2009), 58. Hereafter cited parenthetically in text.

32. Nasser Mufti, *Civilizing War: Imperial Politics and the Poetics of National Rupture* (Evanston: Northwestern University Press, 2017), 128.

33. Ibid., 118.

34. Eloise Knapp Hay, "*Nostromo*," in *The Cambridge Companion to Joseph Conrad*, ed. J. H. Stape (Cambridge: Cambridge University Press, 1996), 88; Peter Lancelot Mallios, "Introduction: Untimely *Nostromo*," *Conradiana* 40, no. 3 (2008): 214.

35. Jakob Lothe, Jeremy Hawthorn, and James Phelan, Introduction to *Joseph Conrad: Voice, Sequence, History, Genre*, ed. Jakob Lothe, Jeremy Hawthorn, and James Phelan (Columbus: Ohio State University Press, 2008), 15; Jacques A. Berthoud and Mara Kalnins, Introduction to *Nostromo: A Tale of the Seaboard*, ed. Jacques A. Berthoud and Mara Kalnins (Oxford: Oxford University Press, 2009), x.

36. Mufti, *Civilizing War*, 131.

37. Hannah Arendt, *On Revolution* (New York: Penguin, 1990), 28.

38. As Said notes, "The immobility that ends *Nostromo* is a sterile calm, as sterile as the future life of the childless Goulds." Edward Said, "The Novel as Beginning Intention," in *Beginnings: Intention and Method* (London: Granta, 1997), 135, quoted in Mufti, *Civilizing War*, 131.

39. Fredric Jameson, *Marxism and Form: Twentieth-Century Dialectical Theories of Literature* (Princeton: Princeton University Press, 1971), 45.

40. Fredric Jameson, "Modernism and Imperialism," in *The Modernist Papers* (London: Verso, 2007), 154.

41. Ibid., 158.

PART II
Form

CHAPTER 4

Fixed Capital and the Flow
Water Power, Steam Power, and *The Mill on the Floss*

Elizabeth Carolyn Miller

George Eliot's novel *The Mill on the Floss* (1860) inhabits dual temporalities at many different levels, formally as well as thematically, as previous critics have discussed. Thinking in world-historical terms, critics such as Suzanne Graver and Nathan K. Hensley have established the novel's investments in epochal shift and in "a theory of time," as Hensley puts it, "with two categories, old and new."[1] John Plotz, in his recent work on the provincial novel, has instead approached the novel's duality in meta-temporal terms, focusing on the famous passage in Eliot's opening chapter where the narrator awakens from a dream-like reverie—"Ah! my arms are really benumbed. I have been pressing my elbows on the arms of my chair, and dreaming that I was standing on the bridge in front of Dorlcote Mill, as it looked one February afternoon many years ago"—to suggest that the strangely doubled temporality inhabited by the novel's narrator, who is half in the past leaning on the bridge near the mill and half in the present seated in an armchair at home, exemplifies "the sort of semi-detached relationship that the reader ... is meant to have to the text itself."[2] Sally Shuttleworth, meanwhile, has drawn attention to the novel's opening sentence—"A wide plain, where the broadening Floss hurries on between its green banks to the sea, and the loving tide, rushing to meet it, checks its passage with an impetuous embrace" (51)—to show how the narrator "disturbs temporal perspective" and to argue, ultimately, in geological-temporal terms that the novel's dual structure is poised between the cyclical and the progressive.[3] *The Mill on the Floss* is temporally double in other, more obvious ways, too: It is a historical novel set in and around 1830 but published in 1860, and the temporal arc of its narrative is curiously bifurcated, with the first part focused on Maggie Tulliver's childhood and the latter part focused on her young adulthood (while remaining, as Deanna Kreisel puts it, "conspicuously silent" about the intervening years in the middle).[4]

My purpose in this essay is to reexamine *The Mill on the Floss*'s temporal structure from the perspective of energy and ecology and to argue that Eliot's well-established interest in dual temporalities and epochal shift extends to a searching and prescient inquiry into the temporality of energy and energy regime transition. For the novel is set at a water-powered mill in the historical moment that saw an unprecedented energy transition in British industry from water power to coal-fired steam power, and it distinguishes between the distinct temporalities of these two energy regimes. This is the moment that saw the birth of what Andreas Malm calls "the fossil economy," when Britain made a "qualitative leap in the manner of coal consumption" that led, more or less directly, to "an economy of self-sustaining growth predicated in the growing consumption of fossil fuels, and therefore generating a sustained growth in emissions of carbon dioxide."[5] Discussion of the possible conversion of Dorlcote Mill from water-powered to steam-powered courses through the novel, and while Eliot was ignorant of the rise in carbon emissions that would accompany the rise of steam, I argue that she recognizes and emphasizes the distinct temporality of a steam-generated economy as opposed to a water-generated one. Time emerges in *The Mill on the Floss* as one vector of human-natural coadaptation, and the novel's temporal doubleness is closely related to its climate and energy imaginary. Along the way, as I make this case, I hope to connect *The Mill on the Floss*'s dual temporality to our present moment of ecological crisis and its demand that we, as critics, shift not so much from an eco-historicism to an eco-presentism, but toward a temporally doubled methodology that inhabits the present and the past dialectically.

A River's a River: Water Power and the Flow

Let us first recall the extent to which water rights, water power, and the transition to coal-fired steam pervade *The Mill on the Floss*. An 1860 review of the novel in *The Spectator* begins with the observation, "The new story by the author of *Adam Bede* is full of power," and this is true in more ways than one.⁶ The opening lines of the novel draw a picture of ships laden with the "dark glitter of coal," moving down the Floss to the town of St. Ogg's (51). We soon learn, in these opening pages, that the father of the novel's heroine, Maggie Tulliver, has a fatal flaw: He is "susceptible in respect of his right to water-power" (55). Dependent on the flow of water to power Dorlcote Mill—a mill that has been in his family "a hundred year and better"—Mr. Tulliver has, at the beginning of the novel, successfully fought off a neighbor's attempt to dam the river, but he is now engaging in a new legal entanglement against Mr. Pivart, a farmer setting up an irrigation scheme farther up the river. "I'll *Pivart* him!" he vows.

Tulliver is convinced that Pivart's irrigation will interfere with his mill. He is convinced of this on the tautological principle "that water was water," a principle he repeats so often as to effectively square its tautology. Water, he says, is "a very particular thing—you can't pick it up with a pitchfork. That's why it's been nuts to Old Harry and the lawyers. It's plain enough what's the rights and the wrongs of water, if you look at it straightforrard; for a river's a river, and if you've got a mill, you must have water to turn it; and it's no use telling me, Pivart's erigation and nonsense won't stop my wheel; I know what belongs to water better than that" (191). Countering Tulliver's irascible and oft-repeated reasoning, Jules Law has examined Riparian doctrine ("the body of laws and precedent concerning water rights") in relation to Eliot's novel to argue that both irrigation technology and the laws governing it were in their infancy at this time, and that it would have been impossible for Tulliver to know how his water-powered mill would be affected by an irrigation system upstream or to predict the outcome of the legal case that eventually ruins him.⁷ Still, it is possible Tulliver may be on to something, for, as W. Jeffrey Bolster demonstrates in *The Mortal Sea: Fishing the Atlantic in the Age of Sail*, nineteenth-century laws around water lagged significantly behind the observations of those who worked the waters in terms of recognizing the need for regulation.⁸

Regardless of whether Pivart's irrigation represents a real threat to his mill, Tulliver's legal woes and his oft-repeated claim that "water is water" point to a fundamental problem at work in the novel and in the broader energy transition happening at the time the novel is set: Water is both spatially and temporally unsuited to privatization, and thus to capitalization on a large scale.⁹ As Tulliver says, "you can't pick it up with a pitchfork." Water power, with wind and solar, sits within the energy category that Andreas Malm calls "the flow." The formal properties of the flow are better suited to collectivization than privatization. Water, for example, is difficult to contain within the bounds of private property: "It respected no deeds or titles, bowed to no monetary transactions; it continued on its course, unmoved by conceptions of private property because it was always in motion."[10] Even if one owns the land on which a stream of water flows, that stream is subject to the actions of other landowners upstream or downstream, which is precisely why the laws around water's use were a matter of dispute at the time the novel is set. Large-scale reservoir schemes developed in the early–nineteenth century, Malm has shown, held the promise of greatly expanding the scale and might of water power in this period, but they would have required a degree of cooperation and coordination among energy users that capitalist competition made unfeasible.[11]

More significant for my argument, water power is also *temporally* unsuited to capitalization at a large scale because it is subject to fluctuations based on the weather and the calendar. Early in the novel, when Maggie goes inside Dorlcote Mill, she hears "the resolute din, the unresting motion of the great stones," as though the mill's power is ceaseless, yet she also senses "the presence of an uncontrollable force," and indeed, the force powering the mill is, in a real way, not fully controllable (72). Water power entails a human harnessing of the river, but dry weather as well as wet weather and storms can impact its capacity. As Jean-Claude Debeir, Jean-Paul Deléage, and Daniel Hémery explain, "if the water was too abundant, the level rose, flooding and immobilizing the wheels; in a drought or in freezing weather, the wheels were again immobilized."[12] And while, as Malm puts it, "traditionally, weak streams during dry summers were no more aberrant or maddening than the fact that grain could not be harvested in midwinter or ploughed in a thunderstorm," such "indulgence toward erratic rivers" had an inverse relationship with the rise of global capitalism. With "the production of commodities for export" and "the maximization of profits through sale on markets detached from the British calendar," the temporal ebb and flow of water power became newly intolerable for 1830s manufacturers, despite the fact that water power was such a cheap and easy means of producing energy in water-rich England.[13]

Debeir, Deléage, and Hémery have shown how, in eighteenth-century England, water power "drove the textile industry to volumes of output previously unknown"—a good reminder that rural capitalism held sway in England long before the rise of steam, as Raymond Williams among others has established.[14] But as with so many issues in the history of capitalist ecology, the question is one of scale. As the water-powered textile industry sought to "meet new needs" and reach new markets, it became more of a problem "when there were freezing temperatures or the stream reached its low-water mark."[15] Coal was not cheap in the time that Eliot's novel is set, as we are reminded when Mrs. Tulliver chides her husband for breaking a large piece of coal in the fireplace: "Mr Tulliver, what can you be thinking of? ... it's very wasteful, breaking the coal, and we've got hardly any large coal left" (289). Though expensive to run, coal-fired steam engines won out over water power because they better suited the abstractions of time and space that accompanied the rise of global capitalism; they offered a release from the temporal oscillations of water, which varied with the seasons and the weather.[16]

Eliot's novel ties water power and the flow closely to the temporal arc of the calendar by situating the devastating flood at the end of the novel—the flood that destroys the mill and kills Maggie and her brother—in the second week of September, around the time of the autumnal equinox. A long-held folk belief in the so-called "equinoctial storm," which held that "a severe storm is due at or near the date of the equinox," was gradually debunked with the improvement of meteorological science in the late nineteenth century.[17] In Eliot's novel, however, the equinoctial storm serves as a climatic and climactic event that ties together weather, time, and water, establishing the temporality of water power as seasonally variable, bound to the calendar, and occasionally catastrophic. Because the equinox marks a moment of equivalence between night and day, the event would seem to evoke temporal balance and stability, but Eliot's depiction of a disastrous equinoctial storm instead suggests that even moments of apparent balance can be moments of historical rupture. This rhymes with Hensley's reading of the novel as registering "a moment of transfigured revolutionary violence (the flood)" within "the midst of a gradualist, organic historical model," but here I read the river's violence not as transfigured revolutionary violence, but, more directly, as a mark of water's potentially calamitous temporality.[18] Such a temporality, characterized by latent disaster, is evident in the narrator's frequent references to the semi-regular flooding of the Floss, euphemistically termed in one passage "the visitation of the floods." Even the town's name refers to a legendary figure, St. Ogg, whose boat was blessed by the Virgin Mary so that "when the floods came, many were saved by reason of that blessing on the boat" (155).

Indeed, the narrator's frequent musings on floods and flooding exemplify another way in which the novel inhabits dual temporalities: The focus of the narrative is on the life of Maggie Tulliver, but the narrator is often given to foreshadowing a future ravaged by flood and disaster. At times, the narrator takes on an apocalyptic, almost post-human perspective, imagining a postdiluvian earth washed clean of human life.[19] At the beginning of Book Four, for example, the narrator recalls experiences on two other rivers—the Rhone and the Rhine—to parallel *The Mill on the Floss*'s account of "this old-fashioned family life on the banks of the Floss":

> Journeying down the Rhone on a summer's day, you have perhaps felt the sunshine made dreary by those ruined villages which stud the banks in certain parts of its course, telling how the swift river once rose, like an angry, destroying god, sweeping down the feeble generations whose breath is in their nostrils, and making their dwellings a desolation. . . . [T]hese dead-tinted, hollow-eyed, angular skeletons of villages on the Rhone oppress me with the feeling that human life—very much of it—is a narrow, ugly, grovelling existence, which even calamity does not elevate, but rather tends to exhibit in all its bare vulgarity of conception; and I have a cruel conviction that the lives these ruins are the traces of, were part of a gross sum of obscure vitality, that will be swept into the same oblivion with the generations of ants and beavers. (292–93)

The passage is one of many in the novel that foreshadow the novel's tragic ending, the flood that sweeps Maggie and Tom Tulliver "into the same oblivion" as the beavers, ants, and villagers destroyed by the floods of the Rhone.[20] And yet, while critics have debated the extent to which the novel sufficiently prepares us for its final, ruinous flood—sometimes arguing, as Jules Law aptly puts it, that "not every foreshadowed ending is an adequately motivated one"—I am more interested in the ways that this foreshadowing requires the narrator to inhabit an eschatological, postdiluvian temporality.[21] Just as the narrator at the beginning of the novel is resting *both* on the bridge overlooking Dorlcote Mill in the past *and* in a chair at home in the present, in this passage the narrator is both on the Rhone surveying the aftermath of catastrophic flooding and on the Floss previewing the flood and destruction to come (and on a third river, the Rhine, in yet another section of the passage). The realistic, quotidian course of Maggie's life on the Floss is set in contrast with a postdiluvian future in which the earth goes on despite the human life that has left it. Human life becomes, from this perspective, but "a narrow, ugly, grovelling existence," "a gross sum of obscure vitality." Such contrasting temporalities could be said to represent human scale versus historical scale, realist time versus the sweep of the epic, or, as Sally Shuttleworth and Jonathan Smith would argue, geological catastrophism versus uniformitarianism.[22] What is particularly notable for my purposes, however, is the extent to which flood and catastrophe are associated with the temporal rhythms of water and water power.

Figure 1. Joseph Mallord William Turner, *Rain, Steam, and Speed—The Great Western Railway* (1844). (Image © The National Gallery, London.)

Every Wheel Double Pace: Steam and Speed

If the equinoctial storm exemplifies the temporality of water power, which is tied to the vagaries of the seasons and the weather and liable to occasional catastrophe, Eliot's novel also identifies and inhabits the temporality of steam. Many critics have associated steam with that general sense of a quickening in the pace of life that we have come to call "modernity." Mary Hammond, for example, refers to the steam engine as "modernity's symbol," and for the editors of a recent special issue on the topic of the energy humanities, steam and speed collapse into "fossil-fueled modernity."[23] One need only look at J. M. W. Turner's painting *Rain, Steam, and Speed* (1844) to see this symbolic relation being visually forged (see Figure 1).

Turner's steam-powered locomotive speeds along a bridge in the face of a driving rain, dwarfing the tiny boat in the river beneath it. The painting depicts the power of steam overcoming the power of water, and its title points to the new temporal regime that accompanied this energy transition to steam.[24]

Eliot's novel likewise ascribes the accelerating tempo of modern life—i.e., modernity—to the rise of steam power. If the wheel of the mill previously defined the rhythm of life on the Floss, time's wheel is quickening its pace under a new energy regime, as Uncle Deane explains to Maggie's brother, Tom: "The world goes on at a smarter pace now than it did when I was a young fellow." Back then, he says, "The looms went slowish, and fashions didn't alter quite so fast: I'd a best suit that lasted me six years. . . . It's this steam, you see, that has made the difference: it drives on every wheel double pace, and the wheel of fortune along with 'em." Deane connects steam's accelerated temporality directly to the rise of global capitalism and its attendant increase in the production of commodities: "Trade, sir, opens a man's eyes. . . . Somebody has said it's a fine thing to make two ears of corn grow where only one grew before; but, sir, it's a fine thing, too, to further the exchange of commodities" (403–4).[25] This sense of a world speeding up on the back of steam-powered capitalism pervades the novel, as when the narrator reflects on the temporal differences between an older economy and a newer one, contrasting "the industrious men of business of a former generation, who made their fortunes slowly," with "these days of rapid moneygetting" (159). And when Bob Jakin asks Tom if he had thought of "making money by trading a bit," Tom is "well pleased with the prospect of a speculation that might change the slow process of addition into multiplication" (328–29). References to speculation, alongside references to "the wheel of fortune" spinning ever more quickly under steam's power, suggest that the temporal profile of steam is accelerated but also risk-prone.

If the steam engine has occasioned a general acceleration of human life and a speeding up of the business of making and getting, it has also, as we know now, simultaneously effected a slowing down in the pace of natural catastrophe. For if the violent flooding of the river Floss is a semi-regular, seasonal event, the violence of the coal-fired steam engine is, as Rob Nixon has eloquently phrased it, a slow violence, "a violence of delayed destruction that is dispersed across time and space, an attritional violence that is typically not viewed as violence at all."[26] This slow violence is, in part, the violence of climate change—a violence that now seems, perhaps, far less slow than it did even when Nixon's book was published in 2011. In fact the slow violence of carbon accumulation is accelerated and not at all slow from the perspective of geological time, but from the very limited perspective of human time, even its acceleration is obscurely gradual.[27] While water power is associated with a catastrophic temporality in *The Mill on the Floss* because of the river's propensity for occasional, destructive flooding, we can now say, then, that an even more disastrous temporality inheres in steam power, albeit one that is slow-building—so slow that it was not grasped for many decades after the transition to steam. Water power's destructive capacity is, we might say, related to capitalism in its pre-steam phase: After all, the reason that Dorlcote Mill is located on the river and is in danger of flooding in the first place is because it relies on the energy provided by the river. But if the mill's water-powered business is acknowledged by Uncle Deane to be "a good one," he considers it only a worthy speculation for his firm if it "might be increased by the addition of steam-power" (270). Here, as elsewhere in the novel, steam produces quantitative effects that ultimately become qualitative by virtue of scale.

Despite the fact that Eliot lacked a full understanding of the accretional effects of steam power across time, her novel does, in its reflections on energy regimes and water power, speak to a fundamental problem therein: the limitations of individual human perspective. Humans are ill-equipped to understand the longer temporal arcs of the energy systems they use, Eliot suggests, because of their short lifespan and transient memories. Indeed, while many critics have fixated on the flood at the end of the novel as a form of unwarranted punishment for Maggie Tulliver, I want to think of it, instead, in the critical terms of the Anthropocene, where we are tasked with rethinking human agency beyond the individual subject. For the catastrophic flood at the end of the novel wreaks havoc on humans like Maggie and Tom in part because of a lapse of human memory across generations. Before the flood begins, the older residents along the Floss have a sense of what is coming, but the young, with their foreshortened memory and experience, fail to take these warnings seriously: "the rains on this lower course of the river had been incessant, so that the old men had shaken their heads and talked of sixty years ago, when the same sort of weather, happening about the equinox, brought on the great floods, which swept the bridge away, and reduced the town to great misery. But the younger generation, who had seen several small floods, thought lightly of these sombre recollections and forebodings" (508). Such a reaction is foreshadowed earlier in the novel when Tom tells Bob Jakin, "there was a big flood once, when the Round Pool was made. *I* know there was, 'cause father says so." Jakin replies, "*I* don't care about a flood comin' . . . I don't mind the water . . . I'd swim—*I* would" (92). Here the italicized "I"s convey the insufficiency of individual human understanding in the face of the long timeline of the floods and the protracted intervals between them. Eliot's notes on flooding, in her research for the novel, circle around this point by documenting accounts of various floods and how they compare to the memories and records of the local communities.[28]

The tragedy at the novel's end thus represents a failure of human collectivity, but also a failure of human cognition—one that the temporal form of *The Mill on the Floss*, I would suggest, is bent on redressing. For at the novel's conclusion, Eliot's narrator is poised between the future and the past, reading them in relation to one another, and such narration seems to call for a reader with a similarly dialectical temporal orientation. Surveying the land that the waters once engulfed from the standpoint of the future, the narrator observes: "Nature repairs her ravages—repairs them with her sunshine, and with human labour. The desolation wrought by that flood, had left little visible trace on the face of the earth, five years after." Moving five years into the future, Eliot's narrator here inhabits a version of the post-Darwinian utopian ecological vision that Benjamin Morgan identifies with William Morris and Samuel Butler, one that foregrounds "the complex interactions between human and nonhuman system," for the repair, the narrator insists, is a joint project of "sunshine" and "human labour."[29] But the narrator goes on to restate the point more precisely: "Nature repairs her ravages—but not all. The uptorn trees are not rooted again; the parted hills are left scarred; if there is a new growth, the trees are not the same as the old, and the hills underneath their green vesture bear the marks of the past rending. *To the eyes that have dwelt on the past, there is no thorough repair*" (522, emphasis added). Far from reiterating a nostalgic vision of pastoral fixity—a stable point of natural homeostasis at some undetermined moment in the past—Eliot describes instead something akin to the shifting baseline discussed by Bolster, the phenomenon in which "each generation imagine[s] that what it saw first was normal, and that subsequent declines were aberrant. But no generation imagined how profound the changes had been prior to their own careers."[30] The point, then, isn't simply that there is no normal in the natural world; it is that the human tendency to assert a normal actually has the effect of masking how profound humanity's impacts on the natural world have been.

Previously, the narrator established that "the mind of St Ogg's did not look extensively before or after. It inherited a long past without thinking of it, and had no eyes for the spirits that walk the streets" (156). But at the end of the novel, the narrator insists that a longer view is possible. There may be "little visible trace" of the flood's destruction, but the trace is there for those who see with an eye to the past as well as the present. Crucially, this temporally dialectical perspective encouraged by the narrator is one that refuses the consolation and recompense of a cyclical temporality. Trees will grow back, *but not the same trees*. Eliot was, as she wrote *The Mill on the Floss*, grappling with the concept of nonprogressive temporality that she encountered in geological and evolutionary theory. *The Origin of Species* appeared during the time that Eliot was writing and, as Smith notes, was read by Eliot and Lewes "within a month of its appearance"; previously, Eliot had also been influenced by the work of geologist Charles Lyell, who denied "that the earth's history was directional" and "that it was cyclical."[31] Shuttleworth has argued that *The Mill on the Floss* is, like *The Origin of Species* itself, "internally divided" between a historical vision of "ordered social growth" and a "simultaneous revelation" of the "contradictions such a historical perspective conceals."[32] Growth was a fantasy that Darwin had difficulty letting go of; Shuttleworth suggests that the same is true for Eliot's novel. Clearly, such an internal division has stakes for the novel's representation of capitalist growth and its attendant energy regimes, too. Indeed, Eliot's probing of energy and capital in the novel is, we might say, the ecological-economic version of her related probing of progress, history, and time.

Conclusion

I have suggested that Eliot's novel inhabits dual temporalities and that it distinguishes between the temporal profiles of two distinct energy regimes—water and steam. Water power's temporal profile is cyclical and seasonal, bound to the calendar and prone to occasional catastrophic flooding; steam power is associated with accelerated modernity and the temporal abstractions of capitalism, but, in a way not fully grasped by Eliot, also effects a peculiarly slow form of accretional devastation that is difficult to witness across human intervals of time. Eliot's novel concludes in a spectacular example of the catastrophic flooding that inheres in water power, and yet the narrator insists on the impossibility of a full recovery from the flood: "Nature repairs her ravages—but not all." The end of the novel thus seems to convey both the cyclical, catastrophic temporality of water power and the irreparable, non-progressive destruction of the steam engine.

Many critics have discussed the need for ecocriticism today to apprehend literature with new, larger time scales, and Timothy Clark has recently suggested more specifically that we read the past with attention to its unintended consequences and with an awareness of our present limited understanding of events and their futurity. This is precisely the mode of reading that the dual temporalities of *The Mill on the Floss* encourage us to take, and that I have aimed to take with this essay. Although Clark finds the realist novel at a disadvantage in grasping the new scale of human agency in the Anthropocene age—"Can the Leviathan of humanity en masse, as a geological force, be represented? No, at least not in the realist mode still dominant in the novel"—he also finds fault with cli-fi novels that "evade most of the present-day moral, political dilemmas" by resorting "to dystopian or apocalyptic scenarios, with a focus on future environmental disaster such as devastating flooding."[33] It is precisely, however, the intertwining of these two narrative modes and two narrative temporalities—the realist mode and the disaster mode—that makes *The Mill on the Floss* unusually resonant for Anthropocene readers. For while the novel is focused on the daily life of Maggie Tulliver, the narrator is forever reminding us of the limited time scale of Maggie's individual human life, and, indeed, of all human life. Such reminders serve to foreshadow the novel's tragic end, but their function is not limited to this narrative effect.

Beyond foreshadowing, the narrator's orientation toward the future positions the story within a much longer temporal scope, one that extends backward as well as forward. St. Ogg's, where the novel is set, is described as "one of those old, old towns which impress one as a continuation and outgrowth of nature, as much as the nests of the bower-birds or the winding galleries of the white ants: a town which carries the traces of its long growth and history like a millennial tree" (153–54). Viewed through a wide enough temporal lens, the passage suggests, the social life of humans becomes simply a part of the natural world, not separate from it.[34] The necessity of such a wide lens in criticism today has been urged most recently by Jason Moore, who writes that "the dualism of Nature/Society . . . is complicit in the violence of modernity at its core" and that "this dualism drips with blood and dirt, from its sixteenth-century origins to capitalism in its twilight." In place of such dualism, he challenges us to "look at the history of modernity as co-produced, *all the way down and through*," to think of "nature" as a relation where "species make environments, and environments make species." While I view coal and the steam engine as more exceptional in the history of capitalist ecology than Moore would allow, his historical emphasis on the ways in which "the work/energy of the web of life is incorporated into the relations of power" nevertheless provides a useful frame for approaching the rival energy regimes and their temporal representation in *The Mill on the Floss*.[35]

In this way, reading backward as well as forward, we might say that the flood at the end of Eliot's novel has the effect of connecting water's formal resistance to capital with the broader resistance of ecology to capitalism that we are forced to confront in the age of climate change. And indeed, while *The Mill on the Floss*'s title suggests that it is a novel about fixed capital, it is also a novel about running water, which Malm has called "the flowing commons" due to its spatial and temporal qualities that resist privatization.[36] I want to conclude, then, with the suggestion that Eliot's novel connects water and steam power with distinct ways of thinking about capital and ecology that are likewise at work in the novel's temporal dialectic. For while *The Mill on the Floss* is a historical novel with a temporally bifurcated narrative and a narrator who incessantly moves from present to future and past to present, it is also a novel with an overriding interest in property, capital, and the creed of economic growth. As a fictional account of the transition from water power to steam, it is presciently attentive to the temporal limitations of individual understanding at this key historical juncture, especially as those limitations relate to the steam-fueled fantasy of permanent growth. At one point in the novel, a character asks Mr. Deane about his "intentions concerning steam," but the narrator of *The Mill on the Floss* has already reminded us of the great gap between human intention and its outcomes in the world: "gentlemen with broad chests and ambitious intentions do sometimes disappoint their friends by failing to carry the world before them" (459, 204). By way of the novel's dual temporal perspectives, the energy transition happening around and through such characters assumes, proleptically, its due historical weight.

Notes

1. Nathan K. Hensley, *Forms of Empire: The Poetics of Victorian Sovereignty* (Oxford: Oxford University Press, 2017), 47. Graver describes Eliot's close study of German social theorist Ferdinand Tönnies, who articulated the social transition from *Gemeinschaft* ("local, organic, agricultural communities that are modeled on the family and rooted in the traditional and sacred") to *Gesellschaft* ("urban, heterogenous, industrial societies that are culturally sophisticated and shaped by the rational pursuit of self-interest in a capitalistic and secular environment") in the era of industrialism. Graver writes, "It is this England—one in which Gemeinschaft was giving way to Gesellschaft— that George Eliot dramatizes" in *The Mill on the Floss* and in other works that take place in the period around the Reform Bill of 1832, including *Middlemarch*. Suzanne Graver, *George Eliot and Community: A Study in Social Theory and Fictional Form* (Berkeley: University of California Press, 1984), 14, 94.

2. George Eliot, *The Mill on the Floss*, 1860 (Peterborough, ON: Broadview, 2007): 52–53. Hereafter cited parenthetically in the text. John Plotz, "The Provincial Novel," *A Companion to the English Novel*, ed. Stephen Arata, Madigan Haley, J. Paul Hunter, and Jennifer Wicke (Chichester: Wiley Blackwell, 2015), 364.

3. Sally Shuttleworth, *George Eliot and Nineteenth-Century Science: The Make-Believe of a Beginning* (Cambridge: Cambridge University Press, 1984), 51, 76.

4. Deanna Kreisel, "Superfluity and Suction: The Problem with Saving in *The Mill on the Floss*," *Novel* (Fall 2001): 92.

5. Andreas Malm, *Fossil Capital: The Rise of Steam Power and the Roots of Global Warming* (London: Verso, 2016), 16, 11.

6. Review of *The Mill on the Floss* in the *Spectator* (April 7, 1860), rpt. in Eliot, *The Mill on the Floss*, Appendix B, 547.

7. Jules Law, *The Social Life of Fluids: Blood, Milk, and Water in the Victorian Novel* (Ithaca, N.Y.: Cornell University Press, 2010), 79.

8. Bolster makes the case for the prescient understanding of water workers over legislators when it came to nineteenth-century laws about water: "It was primarily fishermen, hand-hardened and relatively unlettered, who argued that the watery world they knew firsthand was changing, and not for the better," but their "statements regarding depletion, diminution, and degradation fell on deaf ears." Bolster also notes—in line with Tulliver's suspicions—that "riparian engineering" can affect "the flow and situation of estuarine rivers." W. Jeffrey Bolster, *The Mortal Sea: Fishing the Atlantic in the Age of Sail* (Cambridge: Harvard University Press, 2012), 168, 3.

9. Perhaps this unsuitability has contributed to the enduring assumption, exploded in Bolster's study, that "the ocean exists outside history." Bolster, *Mortal Sea*, 7.

10. Malm, *Fossil Capital*, 117.

11. Ibid., 118.

12. Jean-Claude Debeir, Jean-Paul Deléage, and Daniel Hémery, *In the Servitude of Power: Energy and Civilization through the Ages*, trans. John Barzman (London: Zed, 1991), 75.

13. Malm, *Fossil Capital*, 166.

14. Debeir et al., *Servitude of Power*, 100; Raymond Williams, *The Country and the City* (New York: Oxford, 1973).

15. Debeir et al., *Servitude of Power*, 101.

16. Eliot would have been familiar with these oscillations because she carefully researched river flooding and inundation while preparing to write the novel. George Eliot, *A Writer's Notebook, 1854–1879, and Uncollected Writings*, ed. Joseph Wiesenfarth (Charlottesville: University of Virginia Press, 1981), 36–38.

17. "Equinoctial Storm, or Gale," *The New International Encyclopaedia*, Vol. 7 (New York: Dodd, Mead and Co., 1907), 166.

18. Hensley, *Forms of Empire*, 42.

19. Mary Wilson Carpenter has argued that both *The Mill on the Floss* and *Romola* "are built on subversive appropriations of an apocalyptic structure" of

history. Mary Wilson Carpenter, *George Eliot and the Landscape of Time: Narrative Form and Protestant Apocalyptic History* (Chapel Hill: University of North Carolina Press, 1986), 102.

20. Significantly, the three species destroyed in this passage are all builders—ants, beavers, and people—but their edifices cannot save them from annihilation.

21. Law, *Social Life of Fluids*, 84. Critical debate about the climactic flood is longstanding. Henry James, in an oft-quoted 1866 review, called the novel's conclusion a "defect" of a "serious" order since "the story is told as if it were destined to have, if not a strictly happy termination, at least one within ordinary probabilities." I agree with Hensley that the critical dispute over whether or not the flood is immanent to the novel's development overlooks the novel's larger question of "what counts as 'ordinary.'" It is ordinary for rivers to flood, as Eliot found in her research on flooding; but from the standpoint of abstract capitalist time, this ordinary propensity meant that water power was irregular and erratic. Henry James, "The Novels of George Eliot," *The Atlantic Monthly* (October 1866), rpt. in Eliot, Appendix B: 565–66; Hensley, *Forms of Empire*, 72.

22. Shuttleworth argues that the novel's "concluding flood does not conform to theories of organic evolution but rather to the historical schema of catastrophism, a geological theory discredited by the 1860s that postulated a series of world disasters and successive creations in order to perpetuate the theory of the fixity of species in the face of evidence for their extinction." Smith argues, relatedly but in contrast, that while "the mere existence of a flood suggests catastrophism," in fact the novel "works to place the flood in a local, uniformitarian context." Shuttleworth, *George Eliot*, 53. Jonathan Smith, *Fact and Feeling: Baconian Science and the Nineteenth-Century Literary Imagination* (Madison: University of Wisconsin Press, 1994): 141, 143.

23. Mary Hammond, *Reading, Publishing and the Formation of Literary Taste in England, 1880–1914* (Aldershot: Ashgate, 2006), 51. Brent Ryan Bellamy and Jeff Diamanti, "Editor's Introduction: Envisioning the Energy Humanities," *Reviews in Cultural Theory* 6, no. 3 (2016): 2.

24. While Eliot's novel is not directly concerned with the railway, it is worth noting that both Turner's painting and *The Mill on the Floss* are set in the period when the steam railroad was quite literally changing public notions of speed: The Liverpool & Manchester Railway, Britain's first steam-operated railway line, opened in 1830, and London was connected to the railway network in 1842. Aileen Fyfe, *Steam-Powered Knowledge: William Chambers and the Business of Publishing, 1820–1860* (Chicago: University of Chicago Press, 2012), 102.

25. The "two ears of corn" allusion refers to Jonathan Swift's *Gulliver's Travels*.

26. Rob Nixon, *Slow Violence and the Environmentalism of the Poor* (Cambridge: Harvard University Press, 2011), 2.

27. I am grateful to Jesse Oak Taylor for this observation.

28. Eliot, *Writer's Notebook*, 36–38.

29. Benjamin Morgan, "How We Might Live: Utopian Ecology in William Morris and Samuel Butler," this volume, 141.

30. Bolster, 10.

31. Smith, *Fact and Feeling*, 122, 131.

32. Shuttleworth, *George Eliot*, 62.

33. Timothy Clark, *Ecocriticism on the Edge: The Anthropocene as a Threshold Concept* (London: Bloomsbury, 2015), 73, 79, 78

34. The persistence of this point within Eliot's *oeuvre* positions her, intriguingly, in a post-Darwinian literary trajectory that includes decadence, for Benjamin Morgan has recently made the case for decadence as a "strategic conflation of the natural and artificial," which "represents both a point of origin for and persistent logic of world-ecological discourses that seek to situate economic and political systems more fully within the parameters of natural processes." Benjamin Morgan, "Fin du Globe: On Decadent Planets," *Victorian Studies* 58, no. 4 (2016): 609–35.

35. Jason W. Moore, *Capitalism in the Web of Life: Ecology and the Accumulation of Capital* (London: Verso, 2015), 4, 7, 15.

36. Malm, *Fossil Capital*, 117. Recall that Maggie herself seeks out a common early in the novel when she decides to leave her family and run away to the gypsies, an incident that conveys her yearning for collective union. While Maggie searches for the common, the narrator moves in and out of free indirect discourse; the demarcations between Maggie's perspective and the narrator's are not clearly marked, but rather flow from one to another, and indeed, I would note that the narrator's frequent use of free indirect discourse in this novel formally conveys a hazy distinction between self and other—a "flowing commons" of another sort. Eliot, *Mill on the Floss*, 144.

CHAPTER 5

"Form Against Force"
Sustainability and Organicism in the Work of John Ruskin

Deanna K. Kreisel

> Nature is finite. Capital is premised on the infinite.
>
> —JASON W. MOORE, *Capitalism in the Web of Life*

In a recent manifesto in *PMLA*, environmental-humanities scholar Stacy Alaimo critiques the sustainability discourse of the past few decades, noting that it "echoes the discourse of conservation at the turn of the twentieth century, especially in its tendency to render the lively world a storehouse of supplies for the elite."[1] Alaimo's analysis does not stretch as far back as the nineteenth century, which is where we can find both the origin of the sustainability concept in its contemporary form and the entanglement of sustainability and colonialism implied in Alaimo's remark. While the association of the world-as-storehouse-of-supplies idea and imperial exploitation originated with the mercantilists of the sixteenth century (who saw colonial expansion as a solution to the problem of combining economic growth and national protectionism), it reached its full modern articulation in the nineteenth century, when the explosion of European colonization went hand-in-hand with calls for new global markets to stave off economic stagnation.[2]

In this essay, I will extend, develop, and nuance this critique by tracing the tensions and paradoxes of sustainability discourse back to the nineteenth century, particularly to the work of Victorian art critic, environmental reformer, and heterodox political economist John Ruskin. An essay on Ruskin might, at first blush, seem an odd bedfellow in a critical volume on ecology and empire in the nineteenth century. Ruskin's views on the British Empire were conflicted at best, openly apologist at worst: He supported Thomas Carlyle's Governor Eyre Defence and Aid Committee after the Morant Bay Rebellion (1865) and, incredibly, managed to avoid any mention of slavery in his aestheticizing analysis of J. M. W. Turner's painting, "Slavers Throwing Overboard the Dead and Dying—Typhoon Coming On" (1840)—which Turner himself was inspired to paint after reading an abolitionist tract. Most notoriously, Ruskin openly advocated colonial expansion in his inaugural address for the Slade Professorship of Art at Oxford University (1870): "And this is what [England] must either do, or perish: she must found colonies as fast and as far as she is able, formed of her most energetic and worthiest men;—seizing every piece of fruitful waste ground she can set her foot on, and there teaching these her colonists that their chief virtue is to be fidelity to their country, and that their first aim is to be to advance the power of England by land and sea."[3] The rhetoric of *terra nullius*—(potentially) fruitful *waste* ground—places Ruskin's exhortation firmly in a tradition of apologies for empire stretching back to Thomas More's *Utopia*.[4]

On the other hand, there are several good reasons to turn to Ruskin in order to recover the history of the sustainability idea—not the least of which is that many of the paradoxical characteristics of sustainability that Alaimo highlights are on full display in his work, including the imbrication of economic resource extraction and imperial expansion: "a storehouse of supplies"; "every piece of fruitful waste ground." Ruskin has long been considered one of the founders of the green movement; his screed against industrial pollution, *The Storm-Cloud of the Nineteenth Century* (1884), is the routine starting place for syllabi on Victorian environmentalism, and scholars have been mining his work for precursors to contemporary ecological discourse since the influential studies *Dreams of an English Eden* by Jeffrey Spear (1984) and the edited volume *Ruskin and Environment* (1995)— where Terry Gifford's conclusion explicitly poses the question, "what key concepts [in Ruskin] appear to be useful in our environmental discourse today?"[5] This foundational work has been developed and expanded in more recent scholarship, including Vicky Albritton and Fredrik Albritton Jonsson's *Green Victorians: The Simple Life in Ruskin's Lake District* (2016), Allen MacDuffie's *Victorian Literature, Energy, and the Ecological Imagination* (2014), and recent essays by Sara Atwood, Siobhan Carroll, and Ella Mershon. As MacDuffie eloquently puts it, Ruskin is a touchstone "for a whole host of twentieth- and twenty-first-century ecological economists who seek to put . . . environmental concerns at the center of economic and social analysis."[6]

Yet perhaps the most compelling reason to turn to the work of Ruskin at this particular historical moment is that it can help us better understand our own culture's investment in the sustainability idea. Ruskin is a resolutely heterodox—even iconoclastic—critic who is nevertheless deeply shaped by the values of his time. More importantly, his work combines elements of heterodox political economy with aesthetic and environmental critique. While several recent ecocritics have pinpointed the origins of our current environmental crisis in imperial capitalism—most recently and notably, Jason W. Moore in *Capitalism in the Web of Life* (2015) and Andreas Malm in *Fossil Capital* (2016)—there has been very little work done on the contributions of nineteenth-century political economy, including its heterodox demand-side critics such as Ruskin, to these histories. In this essay, I will consider Ruskin's work—in particular, "The Work of Iron, In Nature, Art, and Policy" (1858) and *The Ethics of the Dust* (1866)—in the context of recent developments in environmental criticism, paying particular attention to three critical nodes: organicism, value, and form.

Form and Force

Ruskin's preoccupation with organicism marks him as an important precursor to contemporary ecocritical discourse. As historian Donald Worster writes about the work of late–twentieth-century environmentalists, "Though they are quick to deny a belief in any nonmaterial or vitalist force in the organism or in the ecosystem, ecologists frequently argue that breaking nature down into its atomistic parts cannot result in a true understanding of the whole."[7] In Ruskin's work, we can see part of the long history of the entanglement of sustainability and organicism, and the roots of current assumptions about the primacy of living matter in the ecosphere— assumptions that have come under recent attack by such critics as Jane Bennett in *Vibrant Matter* and Timothy Morton in his work on dark ecology, and which Ruskin himself implicitly calls into question at key points in his career.

Models of sustainability characterized by metaphors of organic wholeness can be traced back to the eighteenth-century Physiocrats, and deeply influence nineteenth-century economic thinking. The fantasy of a self-contained system where surplus is metabolized in such a way as to nourish and maintain that system is one we find repeatedly throughout Victorian culture. As Catherine Gallagher discusses in *The Body Economic*, both Ruskin and Charles Dickens imagined a self-sustaining sanitation system in which bodily products, including human waste and even corpses, would nourish further production in a closed and infinite cycle of renewal.[8] The idea of the biosphere as a self-sustaining, closed-loop system is one to which writers returned continuously throughout the century; in an 1853 essay entitled "The Circulation of Matter," F. W. Johnston writes, "The same material—the same carbon, for example—circulates over and over again. . . . It forms part of a vegetable to-day—it may be built into the body of a man to-morrow; and, a week hence, it may have passed through another plant into another animal. What is mine this week is yours the next."[9]

This organicist fantasy of sustainability predicated on the perpetual recycling of waste has been inherited by most contemporary mainstream ecological discourse. According to Michelle Niemann, "The environmentalist emphasis on the re-use of waste . . . is based squarely on the organic metaphor and the way the organic self-enclosure of an ecological unit is instituted as an aim."[10] Some recent critics and philosophers have embraced the impossibility of closed-loop organicism and attempted to rehabilitate it as an ethics or poetics of excess: Georges Bataille, Henri Lefebvre, and Gilles Deleuze, among others. As Niemann notes, "Implicit in the aesthetics of excess is the contention that, though closed-circle organicism's containment of decay is seductive, the transgression of that closed circle is, in fact, the organic's condition of possibility. It is by exceeding itself that the organism thrives."[11] As Bataille argues in Volume 1 of *The Accursed Share*, "The living organism, in a situation determined by the play of energy on the surface of the globe, ordinarily receives more energy than is necessary for maintaining life; the excess energy (wealth) can be used for the growth of a system (for example, an organism); if the system can no longer grow, or if the excess cannot be completely absorbed in its growth, it must necessarily be lost without profit; it must be spent, willingly or not, gloriously or catastrophically."[12]

Before diving into the specifics of Ruskin's engagement with organicism and sustainability, it will be helpful to begin with working definitions of both terms. By "organicism" I refer to the ancient doctrine that the universe—and its constituent parts such as ecosystems—are holistic entities that resemble living organisms, particularly in having parts that function in relation to a greater whole. The organicist metaphor can be applied to a wide range of systems, from planets to alluvial plains to corporations to poems.[13] Organicism is not necessarily the same as—and often rejects—

vitalism, which posits the existence of a nonmaterial force or spirit animating living beings, yet there is often a marked slippage in organicist discourse between the *metaphor* of the system as organism and the idea of the system as functionally "alive."[14]

In recent popular ecological discourse, the term "sustainable" operates in lockstep with "organic" (particularly in the latter's current meaning of "all-natural," containing no human-made materials such as synthetic fertilizers and pesticides or genetically modified organisms). "Sustainability," in current usage, can refer both to economic sustainability—in which case it is most often used to refer to sustainable *development*—and environmental sustainability, which has "weak" and "strong" forms.[15] The definition of sustainable development originally formulated at the World Commission on Environment and Development (WCED) in 1987 is startlingly anthropocentric; it denotes "a set of actions to be taken by present persons that will not diminish the prospects of future persons to enjoy levels of consumption, wealth, utility, or welfare comparable to those enjoyed by present persons."[16] Since the WCED report, the concept has been refined to distinguish weak sustainability from strong: weak sustainability refers to the maintenance of a stable stock of total capital, both natural and human-made, and thus assumes that the latter can function as a substitute for the former; strong sustainability argues that human-made capital is not interchangeable with natural resources.[17] As this essay will argue, we can glimpse the outlines of a strong sustainability concept—and its inherent paradoxes—in the mid-career economic and environmental writings of Ruskin, whose vision of a vital natural world thoroughly enmeshes organicism and sustainability.

The term "sustainable" was not used in the sense of minimizing environmental impact until 1976, and was not used to mean "capable of being maintained at a certain level" until 1924.[18] Thus, in order to trace the history of the concept, we have to search for analogous notions operating under other names. The question of how to dispose of economic surplus under capitalism divides the classical Ricardian theorists from the pessimistic heterodox critics, including Malthus and Ruskin. Ricardo and his followers insisted on the benefit of capital accumulation for the growth of the economy, and defended this position with an appeal to Say's Law: "There is no amount of capital which may not be employed in a country, because a demand is only limited by production. No man produces but with a view to consume or sell, and he never sells but with an intention to purchase some other commodity."[19] Therefore, in the long term, it is impossible for there to be overproduction or overaccumulation of capital due to a failure of demand: This principle is the bedrock of *laissez-faire* economic policy. Malthus contravenes this law in his *Principles of Political Economy* (1820) when he argues that "reciprocal demand," or the simultaneous desire of individuals for commodities that can be exchanged for one another, is what determines the value of those commodities, not production or labor costs. Demand is thus no longer a negligible variable that operates in mechanical lockstep with supply; the consequence of this uncoupling is the persistent anxiety that there may be a cataclysmic failure of consumer demand.[20] The strong streak of pessimism in Malthus's work can thus be attributed not only to the theory famously outlined in the *Essay on the Principles of Population* (1798)—that population increases geometrically, while agricultural production "only increases in an arithmetical ratio"—but also to his conviction of human beings' innate laziness and perverse desire to hoard.

Malthus's work is the progenitor of a significant heterodox strain in nineteenth-century political economy, which actively critiqued the fantasy of the self-regulating economy; one of the most important of these critics was Ruskin. In "Ad Valorem," one of the four essays composing *Unto this Last* (1860), his mid-career rebuttal to John Stuart Mill's *Principles of Political Economy* (1848), Ruskin explicitly defines economic value in terms of "life": "*Valor*, from *valere*, to be well or strong;—strong, *in* life (if a man), or valiant; strong, *for* life (if a thing), or valuable. To be 'valuable,' therefore, is to 'avail towards life.'"[21] Ruskin's insistence on vitality as a determinant of value was an important part of his economic heterodoxy; he was openly critical of the dominant labor theory of value found in Adam Smith and Ricardo. J. A. Hobson, one of Ruskin's first exegetes, makes the connection between value and organicism explicit as early as 1898: "Biologists and sociologists correlating the processes of organic life . . . are everywhere engaged in giving intellectual form to a science and art of life such as Mr. Ruskin conceived and foreshadowed in his Political Economy . . . . [H]is 'value' is in substantial conformity to this same scientific purpose."[22] Allen MacDuffie notes that for Hobson, Ruskin's concept of value is essentially identical to the thermodynamic concept of *energy*,[23] a point which Hobson himself makes more or less explicit when he states that by the term "value," Ruskin refers to "the idea of a physical replacement of energy given out in work."[24]

Ruskin's emphasis on the life-sustaining properties of objects of value underpins his general interest in organicism and organic form, the central feature that draws together his diverse writings on architecture, painting, and drawing; political economy and economic theory; and social policy and environmental reform. Yet Ruskin seemingly struggles to define exactly what "life" is. He returns to the question repeatedly throughout his writings, yet two different organic "limit cases" are particular objects of his inquiry: crystals and iron. In *The Ethics of the Dust* (1866), a whimsical dialogue in which Ruskin (the "Old Lecturer") delivers "ten lectures to little housewives on the elements of crystallisation" (the subtitle of the work), Ruskin writes with extraordinary power about the vitality of geological formations:

> Agates, I think, of all stones, confess most of their past history.... Observe, first, you have the whole mass of the rock in motion, either contracting itself, and so gradually widening the cracks; or being compressed, and thereby closing them, and crushing their edges.... Then the veins themselves, when the rock leaves them open by its contraction, act with various power of suction upon its substance.... [Gases] may be supplied in all variation of volume and power from below; or, slowly, by the decomposition of the rocks themselves; and, at changing temperatures, must exert relatively changing forces of decomposition and combination on the walls of the veins they fill; while water, at every degree of heat and pressure ... congeals, and drips, and throbs, and thrills, from crag to crag; and breathes from pulse to pulse of foaming or fiery arteries, whose beating is felt through chains of the great islands of the Indian seas, as your own pulses lift your bracelets.[25]

The "open" veins in the rock exert their terrific "power of suction" in an extraordinary image that makes clear the connection between seemingly "dead" matter and organic life.

Ruskin also makes clear the connection between resource extraction and colonial appropriation. The Indian Ocean had been a site of contention among European colonizing powers since the fifteenth century; by the beginning of the nineteenth, Great Britain had wrested dominance over the region from the Dutch East India Company. The islands Ruskin refers to, most notably Ceylon (now Sri Lanka), remain an important source of gemstones—the Sanskrit name for the Indian Ocean means "jewel mine." In a beautifully involuted metaphor, Ruskin likens the islands themselves to gems: the "pulse" of the earth's geothermic energy lifts the ridges of the islands just as the literal pulses of the girls' veins lift the gemstones of their bracelets. The metaphor thus establishes two different correspondences: between the source of extracted resources (the islands) and the resources themselves (gems); and between the bodies of the girls and the vital "body" of the earth. Ruskin hints at the holistic nature of the global imperial economy, which brings "fruitful waste ground" under cultivation in order to provide products for consumption at the metropole. The organicism of the metaphor—the blurring of the line between living and non-living, and the insistent repetition of images of blood circulation—bolsters an implicit endorsement of such globalism by associating it with vitality, an endorsement that (from a modern perspective) exists in uneasy tension with the more progressive elements of Ruskin's economic critique.

Yet both the lecturer and the schoolgirls he addresses worry throughout the text about whether or not crystals are literally alive. The Lecturer's initial attempt at a resolution yields a definition of life based on *form*:

> I do not think we should use the word "life" of any energy which does not belong to a given form. A seed, or an egg, or a young animal, are properly called 'alive' with respect to the force belonging to those forms, which consistently develops that form, and no other. But the force which crystallises a mineral appears to be chiefly external, and it does not produce an entirely determinate and individual form, limited in size, but only an aggregation.[26]

Ruskin had elaborated on this idea a few years earlier, in volume 5 of *Modern Painters* (1860):

> The mineral crystals group themselves neither in succession, nor in sympathy; but great and small recklessly strive for place, and face or distort each other as they gather into opponent asperities. The confused crowd fills the rock cavity, hanging together in a glittering, yet sordid heap, in which nearly every crystal, owing to their vain contention, is imperfect, or impure. . . . But the order of the leaves is one of soft and subdued concession. Patiently each awaits its appointed time, accepts its prepared place, yields its required observance. Under every oppression of external accident, the group yet follows a law laid down in its own heart.[27]

Yet in the later *Ethics of the Dust*, this attempted definition brings an immediate objection from one of the girls—"But I do not see much difference, that way, between a crystal and a tree"—followed by the Lecturer's response, "Add, then, that the mode of the energy in a living thing implies a continual change in its elements; and a period for its end. So you may define life by its attached negative, death; and still more by its attached positive, birth. But I won't be plagued any more about this, just now; if you choose to think the crystals alive, do, and welcome."²⁸ The felt force of distinction between trees and crystals so apparent in the earlier text is brought sharply into question here.

This moment is repeated later in *Ethics*, when another girl complains, "You always talk as if the crystals were alive; and we never understand how much you are in play, and how much in earnest,"²⁹ to which the Lecturer responds, "Neither do I understand, myself, my dear, how much I am in earnest. The stones puzzle me as much as I puzzle you. They look as if they were alive, and make me speak as if they were; and I do not in the least know how much truth there is in the appearance . . . . What is it to be 'alive'?"³⁰ When pressed, he returns to the question of form in a gnomic utterance: "You may always stand by Form, against Force."³¹ Since external forces—such as geological pressure—can also appear vital, the only way to distinguish the life force from others is that the former "develops that form [of the body in which it inheres] and no other."³² As James Clark Sherburne notes, Ruskin here restates the "Romantic distinction between 'organic' and 'mechanical' form."³³ Coleridge's distinction between mechanical form, which is imposed from without, and organic form, which is generated from within, "leads in turn to the crucial one between 'mechanical' and 'vital' philosophy. The former knows only of 'the relations of unproductive particles to each other.' It can hold good only for a 'dead nature.' In an organic or vital philosophy, elements 'actually interpenetrate' one another to form a living whole."³⁴

As Mershon argues in a recent essay on *The Ethics of the Dust*, for Ruskin the "promise of mineralogical renewal assuages fears about resource depletion."³⁵ The vitalism of crystals in the Ruskin text is thus marked by both "promise" and "fear," as well as being very much of a particular cultural moment: in the 1860s, scientific debates were raging over the organic states of recently discovered liminal forms, and therefore "it was not always clear whether something was dead or alive."³⁶ The implication is that in *Ethics of the Dust* Ruskin tethers the rhetoric of a particular moment in scientific history—a moment of intense debate over the difference between the organic and inorganic—to a broader optimistic argument about what I would term environmental sustainability. For Mershon, *Ethics of the Dust* enacts a fantasy "wherein scant resources are limitlessly recycled and reborn,"³⁷ a fantasy in which "the stakes . . . are nothing less than the expenditure of planetary resources and the annihilation of the human species."³⁸ For Mershon this moment in Ruskin is short-lived, as he soon returns to his usual grumpy predictions about environmental catastrophe, culminating in the apocalyptic vision of *The Storm-Cloud of the Nineteenth*

Century a couple of decades later. Yet I would argue that Ruskin's commitment to the porosity of the life-nonlife border is more intractable, recurring throughout his own work and aligning with a much longer line of thinking throughout European intellectual history.[39] Furthermore, Ruskin's insistent deconstruction of the boundary highlights inherent paradoxes in the sustainability concept as it has developed since the Victorian period.

Rather than simply sidestepping the question of whether crystals are life or non-life, the Lecturer forcefully argues *that the question is undecidable*; in other words, it is not an open-ended question that is currently under debate and potentially will be decided at some future time of greater scientific knowledge, but instead constitutes an incorrectly framed question to begin with. As one of the girls pointedly asks, "It is very delightful to imagine the mountains to be alive; but then,—are they alive?" The Lecturer responds, "You may at least earnestly believe, that the presence of the spirit which culminates in your own life, shows itself in dawning, wherever the dust of the earth begins to assume any orderly and lovely state. You will find it impossible to separate this idea of gradated manifestation from that of the vital power. Things are not either wholly alive, or wholly dead. They are less or more alive."[40] Not only does the Lecturer abandon his earlier gesture toward a definition of life based on form—anywhere "the dust of the earth begins to assume any orderly and lovely state" is a manifestation of the same "spirit" which gives external form to the girls themselves—but he strongly suggests that this "vital power" is present in things that are normally not considered life, and that in fact there is no meaningful distinction between any one object with an ordered form and any other, in terms of vitality. Ruskin has moved here from a working definition of life based on a particular kind of form to a claim that *any* ordered form (particularly if it is aesthetically pleasing) can be considered life.[41]

Sherburne notes that Ruskin's category confusion—or more properly, category refusal—which manifests itself as his "ambiguous use of the word 'vital,'" is "rooted in the Romantic tradition's unwillingness to accept a material organicism."[42] We can see in Ruskin's work the grinding edges of two conceptual tectonic plates: a vitalist (or at least anti-mechanistic) tradition that is the legacy of Romanticism, and post-Darwinian materialism. As George Levine argues, "[D]espite Ruskin's obvious passion for the natural world, manifested ... in the almost mad precision of his observation of the texture of flowers and clouds and of the movements of water and glaciers, he retained ... something of the deep Christian distrust of materiality."[43] In other words (as I would argue), in attempting to finesse his

"Christian distrust of materiality," Ruskin, perhaps inadvertently, reintroduces a different kind of materialism, which manifests itself in an unwillingness to distinguish the organic and the inorganic.[44]

This unwillingness is one which many recent ecocritics share. Bennett, in her book *Vibrant Matter*, poses the salient question, "[C]an nonorganic bodies also have a life? Can materiality itself be vital?"[45] In her inquiry she begins with Gilles Deleuze's short essay, "Immanence: A Life," which posits the existence of "*a* life," an indeterminate vitality or "immanent life that is pure power."[46] Bennett goes on to cite *A Thousand Plateaus*, where Deleuze and Félix Guattari "name metal as the exemplar of a vital materiality; . . . it is metal, bursting with *a* life, that gives rise to 'the prodigious idea of Nonorganic Life.'"[47] For Bennett, this characterization is possible because *a* life is "an activeness that is not quite bodily and not quite spatial, because a body-in-space is only one of its possible modalities. . . . This is the activity of intensities rather than of things with extension in space."[48] Vital materialism depends on a kind of "theory of relativity" as she terms it, wherein "the stones, tables, technologies, words, and edibles that confront us as fixed are mobile, internally heterogeneous materials whose rate of speed and pace of change are slow compared to the duration and velocity of the human bodies participating in and perceiving them. 'Objects' appear as such because their becoming proceeds at a speed or a level below the threshold of human discernment."[49]

Bennett is also concerned with the margins of life. Metals, for example, owe their particular properties, as well as their "metallic vitality," to complex systems of cracks that are caused by loose atoms at the edges of the regular lattice of their structure, which is made up of crystalline "grains": "The line of travel of these cracks is not deterministic but expressive of an emergent causality, whereby grains respond . . . to the idiosyncratic movements of their neighbors . . . in feedback spirals."[50] Ruskin is also interested in the function of dynamic cracks in rock, metal, and crystal, locating in them the geological force that he refuses to differentiate from vitalism: "Observe, first, you have the whole mass of the rock in motion, either contracting itself, and so gradually widening the cracks, or being compressed, and thereby closing them, and crushing their edges,—and, if one part of its substance be softer, at the given temperature, than another, probably squeezing that softer substance out into the veins."[51] Ruskin, like Bennett a century and a half later, is concerned with the life-force of materials that we ordinarily think of as inorganic, in this case considering how cracks and fissures within their crystalline structure form vital systems, liminal spaces at the boundary of life.

Air and Iron

Ruskin's fascination with the vitality of inorganic matter is most clearly demonstrated in an 1858 lecture entitled "The Work of Iron, In Nature, Art, and Policy." He begins the lecture with a striking image of the vitality of iron: "You all probably know that in the mixed air we breathe, the part of it essentially needful to us is called oxygen; and that this substance is to all animals, in the most accurate sense of the word,—breath of life. . . . Now it is this very same air which the iron breathes when it gets rusty. It takes the oxygen from the atmosphere as eagerly as we do."[52] Yet only *rusted* iron demonstrates this extraordinary principle of animation—"iron rusted is Living; but when pure or polished, Dead"[53]—because it is the interaction of iron and air that causes the rusting process; rust is a sign or index of vitality.

Iron also has a crucial aesthetic function: In the form of ferrous oxide, it dyes the veil of nature and the human-made products that are fashioned from it; it brings aesthetic (especially painterly) pleasure in the form of purple hillsides, picturesque red and crimson roof tiles, even a blush upon a cheek. Both the vitality and the beauty of iron are thus contingent upon its interaction with oxygen. Ruskin imagines a poetic merger between air (or spirit) and iron (or body): "[W]hat I wish you to carry clearly away with you is the remembrance that in all these uses the metal would be nothing without the air. The pure metal has no power."[54] All the useful and aesthetic functions of iron are dependent on its being both vital—having a life cycle and the capacity for change that is indicated by rust—and part of a larger system of decay and renewal.[55]

That aesthetic function of iron is, for Ruskin, every bit as important as its purported use value; in fact, more so: "[W]e suppose it to be a great defect in iron that it is subject to rust. . . . On the contrary, the most perfect and useful state of it is that ochreous stain; and therefore it is endowed with so ready a disposition to get itself into that state. It is not a fault in the iron, but a virtue, to be so fond of getting rusted, for in that condition it fulfils its most important functions in the universe."[56] Ruskin insists throughout the lecture that his listeners radically rethink their own relationship to iron and other "natural resources"; we must resist the common way of thinking that, because we "cannot use a rusty knife or razor so well as a polished one," there is something defective in rusty iron.[57] The moment of aesthetic appreciation—the perverse beauty of rusted iron, and the recognition of the role that oxygenated iron plays in the colorful beauty of natural landscape—opens a way toward questioning an instrumental relation to nature.

This is an idea we see throughout Ruskin's work. In a famous passage from *Proserpina*, he writes, "The flower exists for its own sake,—not for the fruit's sake. The production of the fruit is an added honour to it—is a granted consolation to us for its death. But the flower is the end of the seed."[58] Levine notes of this particular passage, "Here, boldly and unapologetically, is the assertion of a value other than use value—an aesthetic value."[59] Levine goes on to insist, however, on the inextricability of aesthetic value and instrumentalism: "[I]n the end for Ruskin everything valuable is valuable insofar as it relates to the human, and the 'non-utilitarian' beauty of the flower is an aspect of the possible moral redemption of man that art (and correct observation) can offer. The whole passage makes clear that the end product is for us."[60] Yet Levine's deconstruction of the distinction between aesthetic and instrumental value does not do full justice to Ruskin's painstaking attempts to limn their differences—attempts that span his entire career. More importantly for the purposes of the current argument, the strong strain of vitalist organicism in Ruskin's writings coexists in uneasy relation to his claims for the priority of human uses of nature. Furthermore, it is precisely this organicism that, as this essay has been attempting to demonstrate, undergirds what we might think of as his sustainability discourse. Ruskin's emphasis on life as the determinant of value, along with his unwillingness to limit the capaciousness of life (rhetorically extending it to iron, crystals, minerals, rust), form the basis of his critique of the instrumental view of natural resources: "It is ourselves who abolish—ourselves who consume: we are the mildew, and the flame."[61]

Where Ruskin's strong sustainability discourse differs from that of contemporary ecologists is in his emphasis on abundance: "[T]he great and only science of Political Economy teaches . . . the service of Wisdom, the lady of Saving, and of eternal fulness; she who has said, 'I will cause those that love me to inherit SUBSTANCE; and I will FILL their treasures.'"[62] As Gill G. Cockram notes, "In Ruskin's economic utopia, the emphasis was on a form of post-capitalist organicism. . . . He had no time for Malthusian notions of a scarcity of resources."[63] Sustainability for Ruskin is not a matter of managing insufficiency, but rather of allocating profusion. As a demand value theorist, his visions of apocalyptic economic failure are characterized by stagnation and gluts brought on by hoarding and insufficient consumption, not by paucity or exhaustion. As David M. Craig argues, Ruskin's focus on "a wealth of the best goods shared among fully developed people" marks him as an heir to the physiocrats; he likewise "always returns to land, agriculture, and food as the 'natural' basis for all wealth."[64]

It is the vision of nature as a storehouse of value that connects Ruskin with some versions of contemporary sustainability theory; the organic metaphor carries with it—however inadvertently or unconsciously—notions of fecundity and infinite renewability. In "Ad Valorem," Ruskin explicitly tethers the notion of renewability to economic *and* aesthetic value: "it will be found at last that all lovely things are also necessary;—the wild flower by the wayside, as well as the tended corn; and the wild birds and creatures of the forest, as well as the tended cattle; because man doth not live by bread only, but also by the desert manna."[65] This utopian vision is underwritten by a fundamental sense of the earth's resources as inexhaustible *because* inexhaustibility is the only version of nature fitted to human needs:

> Men can neither drink steam, nor eat stone. . . . [T]he world cannot become a factory nor a mine. No amount of ingenuity will ever make iron digestible by the million, nor substitute hydrogen for wine. . . . [H]owever the apple of Sodom and the grape of Gomorrah may spread their table for a time with dainties of ashes, and nectar of asps,—so long as men live by bread, the far away valleys must laugh as they are covered with the gold of God, and the shouts of His happy multitudes ring round the winepress and the well.[66]

Notably, these valleys are "far away": the "fruitful waste ground" of the colonies to which Ruskin—and capitalism itself—has constant imaginative recourse. As Moore points out, the "endless frontier strategy of historical capitalism is premised on a vision of the world as interminable: this is the concept of capital and its theology of limitless substitutability."[67] Yet the passage is also complex and contradictory in ways that push beyond the simple "storehouse of resources" argument standard to nineteenth-century political economy. As MacDuffie points out, the passage actually underscores Ruskin's "apprehension of natural limits" because it constitutes a "critique of the idea of substitutability."[68] While the sum total of the stock of natural resources is rhetorically figured as inexhaustible, individual resources are not: Iron is not food. As Moore goes on to argue: "At best, substitutability occurs within definite limits, primarily those of energy flows and the geographical flexibility they offer. The history of capitalism is one of relentless flexibility rather than endless substitutability."[69] The paradox of Ruskin's sustainability discourse—to which the contemporary version is heir—is that its organizing metaphor of organicism, by its very nature, combines anxieties about exhaustibility with utopian visions of infinite plenitude.

Jessica Maynard draws a parallel between Ruskin's emphasis on abundance and the work of twentieth-century anthropologist and economic philosopher Georges Bataille. As I noted earlier, Bataille challenged the recuperative model of closed-loop sustainability by crafting a perverse poetics of excess that celebrates (or at least emphasizes the inevitability of) waste, extravagance, and sumptuary expenditure. For Maynard, this impulse can be traced back to the 1850s; both Bataille and Ruskin, particularly in his discussion of Gothic ornament in *The Stones of Venice*, distinguish between instrumental consumerism "and a second order of consumption that for both is sacrificial, resolutely non-utile in its effects."[70] Most importantly, Maynard underscores the connection I have been insisting on between Ruskin's vitalist organicism and his ethics of consumption, an ethics that emphasizes abundance rather than scarcity: "the sacrificial impulse in his thought might also be related to a dialectical vision of what he calls 'the life of this world.'"[71]

That dialectic is ubiquitous for Ruskin—it is present in both the "changing forces of decomposition and combination" in crystal formation and the bloom and decay of iron. Levine claims that Ruskin's unwillingness to accede to a gross materiality "issues in that astonishing rhetoric that humanizes everything, crystals, leaves, clouds, water."[72] Yet the concomitant of such a rhetoric is a persistent questioning of the boundary not between the human and non-human, but rather between the alive and the inert. Everywhere there is ordered form, for Ruskin, there is a type of "life." While one might argue that for Ruskin it is the uniquely human prerogative to perceive form and thus humanity as the guarantor both of vitality and of an instrumental relation to nature, this essay has attempted to demonstrate that Ruskin's pronouncements on this question are so complex—indeed contradictory—as to indicate a real struggle on his part to distinguish human use value from the abstract value of the natural world. This tendency is perhaps the most troubling aspect of sustainability discourse inherited from Ruskinian political economy—as Alaimo puts it, the way in which that discourse "epitomize[s] distancing epistemologies that render the world as a resource for human use."[73] However, this tendency is the shadowy obverse of a more potentially critical and liberatory strain of thought; Ruskin's refusal of categorical distinctions between living and non-living is also, arguably, the source of the realization that Alaimo describes as "the recognition that one's very self is substantially connected with the world."[74] In both his instrumental and his epiphanic modes, Ruskin is a sustainability theorist *avant la lettre*.

Notes

1. Stacy Alaimo, "Sustainable This, Sustainable That: New Materialisms, Posthumanism, and Unknown Futures," *PMLA* 127, no. 3 (2012): 558.

2. The theory that gluts and stagnation are inevitable without foreign market expansion can be traced back as far as French political economist Jean Charles Léonard Simonde de Sismondi in his *New Principles of Political Economy* (1819). Marxist critics have tended to follow in the footsteps of both Marx and Lenin in agreeing that imperial expansion is a necessary corollary of capitalism. See, for example, the discussion of capitalist/colonial/environmental expansion, particularly the notion of the "frontier," in Jason W. Moore, *Capitalism in the Web of Life: Ecology and the Accumulation of Capital* (New York: Verso, 2015), 63.

3. John Ruskin, *Lectures on Art*, in *Lectures on Art and Aratra Pentelici, With Lectures and Notes on Greek Art and Mythology 1870*, vol. 20 of *The Works of John Ruskin*, ed. E. T. Cook and Alexander Wedderburn (London: George Allen, 1905), 42.

4. For a discussion of the source of the *terra nullius* concept in More, see Carol Pateman, "The Settler Contract," in *Contract and Domination*, ed. Carole Pateman and Charles Mills (Malden, Mass.: Polity Press, 2007), 35–78.

5. Terry Gifford, "Conclusion," in *Ruskin and Environment: The Storm-Cloud of the Nineteenth Century*, ed. Michael Wheeler (Manchester: Manchester University Press, 1995), 188.

6. Allen MacDuffie, *Victorian Literature, Energy, and the Ecological Imagination* (Cambridge: Cambridge University Press, 2014), 169.

7. Donald Worster, *Nature's Economy: A History of Ecological Ideas*, 2nd ed. (Cambridge: Cambridge University Press, 1994), 21–22.

8. Catherine Gallagher, The Body Economic: Life, Death, and Sensation in Political Economy and the Victorian Novel (Princeton: Princeton University Press, 2006), 100–7.

9. James F. W. Johnston, "The Circulation of Matter," *Blackwood's Edinburgh Magazine* 73 (1853): 552.

10. Michelle Niemann, "Rethinking Organic Metaphors in Poetry and Ecology: Rhizomes and Detritus Words in Oni Buchanan's 'Mandrake Vehicles,'" *Journal of Modern Literature* 35, no. 1 (2011): 213.

11. Niemann, "Rethinking," 112.

12. Georges Bataille, *The Accursed Share*, vol. 1, trans. Robert Hurley (New York: Zone Books, 1991), 21.

13. The legacy of organicism in aesthetic criticism is a long and storied one: The standard narrative is that the ideal is born of Aristotle's *Poetics*, is refined and expanded by Coleridge and elaborated by Wordsworth and fellow

Romantics (among whom we may include Thomas Carlyle along with Ruskin), reaches its apotheosis in the criticism of T. S. Eliot and the New Critics, and beginning with Raymond Williams has been under steady attack ever since. The crude form of this argument is that literature that celebrates organicism in nature or society, or takes organic structure as its *summum bonum*, is inherently conservative and politically oppressive.

14. For example, the work of Prussian naturalist Alexander von Humboldt, who theorized that there is a process of "coevolution" between living organisms and the earth's crust and climate. Humboldt's theory anticipates much later ecological and conservationist discourse, such as some elements of living systems theory, or James Lovelock's "Gaia model" of the 1970s, which argues that the entire Earth functions as a self-regulating, living system.

15. It is important to distinguish development from growth: the former refers to increases in overall well-being, which can be brought about through such measures as redistribution and education; the latter refers to an (inflation-adjusted) increase in the total value of the goods and services produced by a particular economy.

16. Daniel W. Bromley, "Sustainability," in *The New Palgrave Dictionary of Economics*, 2nd ed., ed. Steven N. Durlauf and Lawrence E. Blume (Palgrave Macmillan, 2008), http://www.dictionaryofeconomics.com/article?id=pde2008_S000482.

17. The concept of weak sustainability is articulated by Hartwick's Rule. See John Hartwick, "Substitution among Exhaustible Resources and Intergenerational Equity," *Review of Economic Studies* 45, no. 2 (1978): 347–54; and Robert M. Solow, "On the Intergenerational Allocation of Natural Resources," *Scandinavian Journal of Economics* 88, no. 1 (1986): 141–49.

18. *Oxford English Dictionary*, s.v. "sustainable," accessed March 12, 2017, http://www.oed.com.

19. David Ricardo, *The Principles of Political Economy and Taxation* (London: Dent, 1973), 192–93.

20. I thus read Malthus as an important precursor to marginal utility theory at the end of the century. For a fuller discussion, see the introduction to Deanna K. Kreisel, *Economic Woman: Demand, Gender, and Narrative Closure in Eliot and Hardy* (Toronto: University of Toronto Press, 2012), 3–24.

21. Ruskin, *Unto This Last*, in *Unto This Last, Munera Pulvaris, Time and Tide, With Other Writings on Political Economy 1860–1873*, vol. 17 of *The Works of John Ruskin*, ed. E. T. Cook and Alexander Wedderburn (London: George Allen, 1905), 84.

22. J. A. Hobson, *John Ruskin, Social Reformer* (London: James Nisbit, 1898), 88.

23. MacDuffie, *Victorian*, 147.
24. Hobson, *John Ruskin*, 139.
25. Ruskin, *The Ethics of the Dust*, in *Sesame and Lilies, The Ethics of the Dust, The Crown of Wild Olive, With Letters on Public Affairs 1859–1866*, vol. 18 of *The Works of John Ruskin*, ed. E. T. Cook and Alexander Wedderburn (London: George Allen, 1905), 333.
26. Ibid., 238–39.
27. Ruskin, *Modern Painters Volume V*, vol. 7 of *The Works of John Ruskin*, ed. E. T. Cook and Alexander Wedderburn (London: George Allen, 1905), 49–50.
28. Ruskin, *Ethics*, 239.
29. Ibid., 340–41.
30. Ibid., 341.
31. Ibid.
32. Ibid., 239.
33. James Clark Sherburne, *John Ruskin, or the Ambiguities of Abundance: A Study in Social and Economic Criticism* (Cambridge: Harvard University Press, 1972), 8.
34. Ibid., 8–9.
35. Ella Mershon, "Ruskin's Dust," *Victorian Studies* 58, no. 3 (2016): 466.
36. Ibid., 476.
37. Ibid., 485.
38. Ibid., 479.
39. For histories of the debate over the definition of life, see S. Tirard, M. Morange, and A. Lazcano, "The Definition of Life: A Brief History of an Elusive Scientific Endeavor," *Astrobiology* 10, no. 10 (2010): 1003–9; and G. R. Welsh and J. S. Clegg, "From Protoplasmic Theory to Cellular Systems Biology: A 150-Year Reflection," *American Journal of Physiology—Cell Physiology* 298, no. 6 (2010): C1280–C1290.
40. Ruskin, *Ethics*, 346. This is an idea Ruskin had toyed with as early as *The Stones of Venice* (1851): "Nothing that lives is, or can be, rigidly perfect; part of it is decaying, part nascent. The foxglove blossom,—a third part bud, a third part past, a third part in full bloom,—is a type of the life of this world." *The Stones of Venice Volume II: The Sea-Stories*, vol. 10 of *The Works of John Ruskin*, ed. E. T. Cook and Alexander Wedderburn (London: George Allen, 1905), 203.
41. Andrea Pinotti locates a similar impulse toward vitalism in *The Elements of Drawing* (1857), in "Gothic as Leaf, Gothic as Crystal: John Ruskin and Wilhelm Morringer," in *Ruskin and Modernism*, ed. Giovanni Cianci and Peter Nicholls (New York: Palgrave 2001), 24.
42. Sherburne, *John Ruskin*, 127.

43. George Levine, "Ruskin, Darwin, and the Matter of Matter," *Nineteenth-Century Prose* 35 (2008): 236–37.

44. For a very helpful reading of another passage in *The Ethics of the Dust* that reaches a similar conclusion, see Sharon Aronofsky Weltman, *Ruskin's Mythic Queen: Gender Subversion in Victorian Culture* (Athens: Ohio University Press, 1998), 141.

45. Jane Bennett, *Vibrant Matter: A Political Economy of Things* (Durham, N.C.: Duke University Press, 2010), 53.

46. Quoted in Bennett, *Vibrant*, 53.

47. Ibid., 55.

48. Ibid.

49. Ibid., 57–58.

50. Ibid., 59.

51. Ruskin, *Ethics*, 333.

52. Ruskin, "The Work of Iron, In Nature, Art, and Policy," in *"A Joy Forever" and The Two Paths, With Letters on The Oxford Museum and Various Addresses 1856–1860*, vol. 16 of *The Works of John Ruskin*, ed. E. T. Cook and Alexander Wedderburn (London: George Allen, 1905), 377.

53. Ibid., 376–77.

54. Ibid., 385.

55. This restless principle of vitality can be traced back to Lyell's *Principles of Geology*: "The renovating as well as the destroying causes are unceasingly at work, the repair of land being as constant as its decay." Quoted in Andrea Charise, "G. H. Lewes and the Impossible Classification of Organic Life," *Victorian Studies* 57, no. 3 (2015): 381.

56. Ruskin, "The Work," 376.

57. Ibid.

58. Ruskin, *Proserpina*, in *Love's Meinie; and Proserpina*, vol. 25 of *The Works of John Ruskin*, ed. E. T. Cook and Alexander Wedderburn (London: George Allen, 1906), 350.

59. Levine, "Ruskin," 239.

60. Ibid.

61. Ruskin, "A Joy Forever," in *"A Joy Forever" and The Two Paths, With Letters on The Oxford Museum and Various Addresses 1856–1860*, vol. 16 of *The Works of John Ruskin*, ed. E. T. Cook and Alexander Wedderburn (London: George Allen, 1905), 64.

62. Ruskin, *Unto This Last*, 85.

63. Gill G. Cockram, *Ruskin and Social Reform: Ethics and Economics in the Victorian Age* (New York: Tauris Academic Studies, 2007), 42.

64. David M. Craig, *John Ruskin and the Ethics of Consumption* (Charlottesville: University of Virginia Press, 2006), 270.

65. Ruskin, *Unto This Last*, 111.
66. Ibid., 110.
67. Moore, *Capitalism*, 66.
68. MacDuffie, *Victorian*, 154.
69. Moore, *Capitalism*, 66.
70. Jessica Maynard, "Architectures of Sacrifice: Ruskin, Bataille, and the Resistance to Utility," *Mosaic: A Journal for the Interdisciplinary Study of Literature* 39, no. 1 (2006): 115–30.
71. Ibid.
72. Levine, "Ruskin," 237.
73. Alaimo, "Sustainable," 563.
74. Ibid., 561. Moore also makes explicit the necessity of this reconceptualization: "We must have a way of naming—and building the conversation through—the relation of life-making. . . . So we begin with an open conception of life-making, one that views the boundaries of the organic and inorganic as ever-shifting." *Capitalism*, 7.

CHAPTER 6

Mapping the "Invisible Region, Far Away" in *Dombey and Son*

Adam Grener

One of the yields of the ecological turn in literary studies has been renewed attention to the novel form's capacity to represent vast and aggregate social processes, both in terms of their enabling conditions and in terms of their environmental impact. Recent work by Allen MacDuffie and Jesse Oak Taylor, for instance, has shown how the novel form might be productively understood in relation to thermodynamic systems and climate models, respectively. Revealing the ecological dimensions of the Victorian literary imagination has offered new ways of understanding how the novel represents "character, event, and environment [as] mutually shaping, reciprocally expressive, and systematically interconnected."[1] This methodological turn has promoted a shift in critical attitudes toward realism as well, from an emphasis on realism's inherent limitations in representing the vast global network that subtends the "reality" it concerns itself with toward an exploration of how its attention to a circumscribed milieu can nevertheless register the complex dynamics of capitalism.

This essay extends this work and refines its theoretical coordinates by examining the relationship between empire and ecology in Charles Dickens's *Dombey and Son* (1846–48), emphasizing in particular the important role the atmosphere plays in mapping the connections between domestic, national, and imperial spaces in the novel. With a full title of *Dealings with the Firm of Dombey and Son, Wholesale, Retail, and for Exportation*, Dickens's novel is concerned with the alignment (and misalignment) of the domestic and economic spheres as it enacts the deflation of Dombey's pride and the collapse of his firm before his eventual redemption through his recognition of his daughter Florence. Although the novel embraces an expansive vision of social reform, the Empire seems peculiarly excluded from a totalizing view that links the domestic and the economic. As critics such as Suvendrini Perera have shown, *Dombey* is a "parable of mercantile capitalism" that is "predicated on an economy of empire."[2] The absence of these spaces from the novel's explicit narration seems indicative of its inability to fully account for the activities of that economy. However, attention to ecological details—and the novel's preoccupation with atmospheric and meteorological phenomena in particular—reveals how the logic of the novel's reforming vision nevertheless incorporates this "invisible region, far away."[3] Jason Moore's *Capitalism in the Web of Life* (2015) has prompted us to rethink the Cartesian binary of nature/society that has structured both Green Thought and analyses of capitalist accumulation, reminding us that capitalism (and human activity more broadly) does not just act upon nature but is embedded within a web of life that shapes the contours of that activity and is in turn reciprocally shaped by it.[4] Nature might appear simply as mere backdrop or an object of capitalist plunder in the novel—the "raw materials" Mr. Baps repeatedly mentions that "came into your port in return for your drain of gold" (221)—but following Moore's lead reveals the imbrication of empire and ecology in Dickens's novelistic form. The novel's preoccupation with weather and the meteorological, in other words, is not just a byproduct of the project of empire—weather is not simply something that must be navigated in the exploitation of the regions of empire. Instead, weather imposes limits, and, as a system, makes otherwise inaccessible regions obliquely legible in the metropole. This in turn makes possible the inclusion of imperial spaces in the novel's vision of reform.

 Reading the ecological in *Dombey and Son* back against the Empire in this way also spells out the representational logic by which the form of the Victorian novel is able to map networks of systematic interconnection.[5] Its investment in the dynamics of the atmosphere and weather is particularly poignant in this regard since meteorology was emerging as a science increasingly attuned to the ways in which local conditions were part of a global system and affected by events in regions far away. While the Empire itself remains an "invisible region" unrepresented in the pages of the novel,

Dickens's reforming vision invokes a purported omniscience that can represent connections between individuals and the social whole, between localized actions and their distant ramifications. The mechanics of this point of view are presented most explicitly in the famous invocation in Chapter 47 of an Asmodean spirit to "take the house-tops off . . . and show a Christian people what dark shapes issue from amidst their homes" so as to "rous[e] some who never have looked out up on the world of human life around them, to a knowledge of their own relation to it" (702). However, the novel's seemingly contradictory optics—simultaneously invoking the fantasy of complete visibility while constantly appealing to an invisible region in metaphysical but also geographical terms—highlight the constitutive tension of a realist aesthetic committed to the particularity of local environments situated within a global network. *Dombey*'s reforming vision dramatizes how the capacity of the novel form to represent ecological "models" and "systems" is predicated not only upon its capacity to trace metonymic connections but also upon the assumption of a totality of interconnections that can be imagined but not fully represented. The invisibility of imperial space is not a failure of the novel's totalizing vision; rather, its peculiar and partial visibility within the circumscribed view of the novel through the mediating space of the atmosphere points to the combination of particularity and abstraction required to represent systemic interconnection.

Sea: The Limits of Metonymy

Dombey and Son has been regarded as a turning point in Dickens's career where he begins to harness the formal capacities of the multi-plot serial novel in the systematic representation of a rapidly changing Victorian world. Dickens had claimed in the Preface to *Martin Chuzzlewit* (1843–44) that he had "endeavoured . . . to resist the temptation of the current Monthly Number, and to keep a steadier eye upon the general purpose and design," but it is only in this next novel where the effects of this steadier eye become more clearly evident.[6] *Dombey*'s formal coherence resides not only in its controlled negotiation of monthly numbers and novelistic whole (evidenced by Dickens's first set of robust working notes),[7] but also in its tactical management of—and movement between—the various layers of social reality it represents, coordinating psychological and domestic interiors, diverse regions of the urban milieu, and the networks of exchange and transport that traverse the nation. Steven Marcus, for instance, suggests that *Dombey and Son* is not only the first of Dickens's works that "might be thought of as a domestic novel," but also one that presents a "singleness of purpose" as it "undertakes a comprehensive, unified presentation of social life by depicting how an abstract principle conditions all experience."[8] That principle, as Marcus and others have noted, is change, which the novel engages through its two dominant images of the railroad and the sea. The railroad possesses the power to both contract the temporal and spatial dimensions of the nation and to reshape its landscape in the image of progress. The remaking of Staggs's Gardens in Chapter 6 is cataclysmic—"the first shock of a great earthquake"—yet from "this dire disorder" flows the "mighty course of civilisation and improvement" (78–79). However, if the novel seems to embrace the energies of industrial progress, it does so primarily at the level of the collective whole. Dombey's solipsistic vision of an "earth made for Dombey and Son to trade in, and the sun and moon . . . made to give them light" is the novel's primary target of reform, precisely because it entails a collapse into Dombey (and Son) of the collective, global resources harnessed by his firm (12). If Dombey's initial worldview represents one end of a spectrum—the globe as material platform for self-realization—then the reforming vision of Chapter 47 constitutes the other end. Rather than an individual viewing the world as "a system of which they were the centre," the removal of house-tops brings about a finer calibration of the individual's understanding of their position within that system (12). One dominant thread of the novel's unfolding, then, is the movement from one pole to the other, not only through its thematic dynamics but also through the production of metonymic networks that provide readers and characters alike with the cognitive armature needed to situate individual actors and actions within that global system.

Within this production of a totalizing vision, however, the Empire is pointedly absent, or rather present only insofar as its artifacts infiltrate the mapped regions of the nation itself. Indeed, the novel is littered with objects from and references to the spaces that are the foundation of both Dombey's business and the broader capitalist economy that fuels the social transformations it so assiduously charts. Conduits to the farthest imperial reaches are located within the novel's topography: Dombey's offices and Solomon Gills's Wooden Midshipman are "just round the corner" from the East India House, and Captain Cuttle lives on a canal near the India Docks (46). Characters travel to and from these spaces, engaged in the economic transactions of empire: Walter Gay travels to Barbados, returns on a "China trader," and departs to China with Florence late in the novel (863); Master Blitherstone, "born beneath some Bengal star of ill-omen" is "on ship-board, bound for Bengal" at the novel's end (629, 914); Alice

Marwood returns from where "convicts go," "beyond seas" (525, 527); and Sol Gills navigates the Caribbean—from Barbados to Jamaica to Demerara—in his relentless search for Walter before returning to London. There are even frequent discussions of the processes of resource extraction and commodity exchange through which these spaces are controlled and exploited. These include Walter's promise to send "shiploads" of "lively turtles, and limes for Captain Cuttle's punch, and preserves for [Sol] on Sundays," remembrances of Mr. Pipchin, who died "pumping water out of Peruvian mines," Mr. Baps's speculations about the dynamics of mercantilism, and mentions of "tax-gatherer[s] in the British Dominions—that wide-spread territory on which the sun never sets, and where the tax-gatherer never goes to bed" (288, 115, 345). And then, of course, there is Major Bagstock's "Native," who is the recipient of his constant abuse and whose silent suffering embodies the violence implicit in the operations of empire.

Yet while suggestive references to the Empire are constantly present, the spaces themselves are not. As Elaine Freedgood notes more generally of mid-century novels, "There is virtually no elaboration of what was going on 'out there' in the colonies that might be affecting, or more accurately underwriting, the domestic worlds of novels like . . . *Dombey and Son*."[9] The classification of Bagstock's servant foregrounds the stark contrast between the novel's mapping of the domestic "labyrinth of narrow streets and lanes and alleys" and its reduction of the Empire to an undifferentiated "invisible region, far away" (90). Miss Tox, we learn, is "content to classify [Bagstock's servant] as a 'native,' without connecting him with any geographical idea whatever" (102). Aside from a brief portrait of Carker and Edith's apartment in Dijon, the narrative perspective—so characteristic of a Dickensian "omniscience" that is anything but—never represents spaces beyond the circumscribed geographical border of the nation itself. When characters go to sea—as Walter and Sol do for hundreds of pages—they become invisible, with characters and readers alike left to speculate upon their position and well-being. If then, as Audrey Jaffe has argued, Dickensian omniscience "creates its characteristic effects precisely by establishing and then violating . . . boundaries,"[10] barriers such as the house-tops that materially separate private and public spaces, then it seems that the boundary between the metropole and an empire that materially underwrites its development remains inviolate in *Dombey and Son*. The more we look for metonymic chains of association to link the metropole to its imperial network, the more immaterial the Empire seems to become. Thus, the "objects" adjacent to Dombey's offices and the Wooden Midshipman offer "hints of adventurous and romantic story," while the nearby East India House "teem[s] with suggestions of precious stuffs and stones, tigers, elephants, howdahs, hookahs, umbrellas, palm trees, palanquins, and gorgeous princes of a brown complexion"—the emphasis being here on "hints" and "suggestions" (46). With this formal asymmetry, *Dombey and Son* offers another instance of what Ayşe Çelikkol identifies as a tension between "circulation and enclosure" that reflects the structural tension between capitalism and the nation-state in the era of free trade, whereby "the former needs capital to move without barriers, the latter needs to present itself as a stable, closed community."[11] It also reflects the representational problems that Fredric Jameson has argued accompany the move from market to monopoly capital, whereby "the truth of [the] limited daily experience of London lies, rather, in India or Jamaica or Hong Kong . . . bound up with the whole colonial system of the British Empire that determines the very quality of the individual's subjective life [but whose] structural coordinates are no longer accessible to immediate lived experience."[12]

Garrett Stewart, in the most theoretically sophisticated account of empire in *Dombey and Son*, has pursued this idea to highlight the ideological work the sea performs in relation to empire. The sea is not only, alongside the railroad, *Dombey*'s controlling image, but it is also the hinge in its construction of space—the sea is that which both links and separates the metropole from its imperial sources of wealth. Stewart shows how the rhetorical figure of syllepsis—with a doubleness that "yok[es] unlike things together by a logic somewhere in the middle zone between metonymy and metaphor"—operates as a "metatrope" in the novel, linking the material and spiritual to enact, through the mechanics of the novel's language, the ideological legitimation of empire.[13] The linguistic device that couples the literal and the figurative—for example, little Paul "borne by Fate and Richards" (80), or Mr. Dombey "stiff with starch and arrogance" (110)—can be "extrapolated to a formal principle" that helps us to understand the workings of the novel's "explicit imperial thematic of divided vision, now terrestrial (territorial), now transcendental."[14] Through this process the Empire becomes associated with death, but this coupling collapses the vast colonial project of domination into a redemptive ethic of individual courage through Walter Gay's success. In the yoking of material geographical horizons to death, Stewart argues, the "tenuous ligatures of colonial interdependency come to us refigured as immaterial, distanced, disembodied, impersonal, abstracted to all that remains unseen to be believed, believed in as British fortitude rather than exploitation."[15] Rather than incorporating the vast systems of empire into a collective, unified consciousness, the novel's form elides them in the individual actions that constitute the bedrock of narrative representation. For Stewart, the questions of colonial dominance and exploitation are reduced to an ethics of individual valor and fortitude. If the novel attempts to regulate a global capitalist economy according to the values of familial love, its inability to do so is belied by its failure to extend this model abroad.

Sky: The Precipitation of the Invisible

In *Dombey and Son*, however, the sea is also associated with another set of tropes that provide an alternative way of understanding the novel's engagement with the formal and figural problem of the global system. The novel's deployment of atmospheric and meteorological conditions keeps the "ligatures of colonial interdependency" in view, but does so in a manner that reflects the individual's limited capacity to conceptualize a global system that remains (as Jameson reminds us) inaccessible to lived experience. In the same way that Mary Favret shows how the weather emerged in the Romantic period as a metaphor for reading the climate of war, the sky in *Dombey and Son* becomes a space where the effects of distance are registered.[16] The atmosphere—and bad weather in particular—is inextricably associated with the sea in the novel, the space of romance and adventure that serves as the primary arena of "British fortitude." If the sea's metaphysical associations threaten to dematerialize empire and elide its structural relation to domestic space, its physical expanse keeps questions of empire in view. Early in the novel, for instance, Walter's future prospects are both speculated upon and celebrated over a bottle of Sol's Madeira in the back parlour of the Wooden Midshipman. The ritual both fuels the romance of "the marvellous and adventurous" and points toward Walter's impending nautical misadventures (55). The wine has "'been to the East Indies and back . . . and has been once around the world,'" enduring "'the roaring winds, and rolling seas'" and "'the thunder, lightning, rain, hail, [and] storms of all kinds.'" (53–54). After Walter departs and his safe arrival in Barbados remains uncertain, Sol's uneasiness leads him to think "of raging seas, foundering ships, drowning men, [and] an ancient bottle of Madeira never brought to light" (338). Indeed, the Madeira's ritualistic status triangulates individual ambition, the spatial logic of empire, and a natural world that shapes those activities.[17]

Bad weather might be seen as the obverse of the fortitude that ideologically underpins the project of empire—it is conjoined to the activities that delimit the "geopolitical horizons"[18] of the novel, but works against that delimitation by bringing into view the conditions that shape economic activity. The shipwreck of the *Son and Heir* in "most uncommon bad weather" serves most immediately, on the level of plot, as a threat to the safety and prospects of Walter Gay, yet it also embodies the "unsuccessful ventures" that bring down Dombey's firm and are endemic to the capitalist system more broadly (363, 877). A more complex engagement with the dynamics of empire becomes apparent, then, if we think of nature's relationship to capital not simply as mines to be pumped and raw materials to be imported but rather as an encompassing web that structures the project of empire. As Jason Moore has suggested, "civilizations . . . do not 'interact' with nature as resource (or garbage can); they develop *through* nature-as-matrix."[19] Rather than seeing nature and society as binary entities, focusing on the ways in which society acts upon an inert or passive nature, Moore's effort to think beyond this binary opens avenues for analyzing how capitalist accumulation organizes nature in historically specific configurations and is at the same time co-produced with that nature. While Moore's analysis itself is focused on understanding the various configurations and appropriations of "Cheap Nature," his methodology also facilitates a broader rethinking of the ecological dimension of novelistic form. In the case of *Dombey and Son*, it shows how nature imposes particular limits to imperial domination, thus expanding the framework of the novel's engagement with empire and refining our understanding of the implications of its nautical and meteorological motifs.

It also points to how the narratively invisible regions of empire are nevertheless mapped in the novel. The recalibration of domestic and economic values that is accomplished through Dombey's bankruptcy and reform implies a techno-scientific mastery of these regions through the instrumental knowledge embodied by Sol Gills. His Wooden Midshipman, which serves as both his home and place of business, is economically stagnant at the outset of the novel. His stock in trade consists of "chronometers, barometers, telescopes, compasses, charts, maps, sextants, quadrants, and specimens of every kind of instrument used in the working of a ship's course" (46–47). Yet these instruments, as Michael Klotz has suggested, are merely "decorative" pieces rather than commodities.[20] This detachment from the realm of exchange enables the Wooden Midshipman to serve as an idealized domestic space for much of the novel—it is where Florence goes following her flight from Dombey's house. Walter's departure and subsequent wreck aboard the *Son and Heir*, however, prompt the remobilization of these instruments. Upon his departure to Barbados, Sol and Captain Cuttle "kept [the] reckoning [of the ship] in the little back parlour and worked out her course, with the chart spread before them" (300). As fears of disaster mount in the absence of news of the ship's arrival, Sol amplifies his nautical charting:

> On the table, and about the room, were the charts and maps on which the heavy-hearted Instrument-maker had again and again tracked the missing vessel across the sea, and on which, with a pair of compasses that he still had in his hand, he had been measuring, a minute before, how far she must have driven, to have driven here or there. (369)

Sol eventually sets off in search of his nephew, and although he does not find him, he makes use of his dormant knowledge in his "wanderings" about the Caribbean and return to London, gaining passage on ships were he is "able, now and then, to do a little in return, in [his] own craft" (863). Although not explicitly related, the rising fortunes of Sol's business are linked to this deployment of the technical knowledge he possesses. Not only does the Wooden Midshipman do its "usual easy trade" by the novel's close, but "some of Mr Gills's old investments are coming out wonderfully well; and . . . instead of being behind the times in those respects, as he supposed, he was in truth, a little before it" (943).

The redeployment of Sol's instrumental knowledge might seem to imply a more efficient and effective mastery of the spaces of empire, but it also makes possible their incorporation into the novel's totalizing vision of reform. Even though the economic rejuvenation of the Wooden Shipman challenges its status as domestic sanctuary, its entry into the system of exchange enables the exportation of its values—it is a space, after all, that is first introduced as a "snug, sea-going, ship-shape concern, wanting only good sea-room, in the event of an unexpected launch, to work its way securely, to any desert island in the world" (47). If it is the weather that presents the most explicit limit to the capitalist system in the novel, weather also offers the primary structure for how the novel understands the interconnections between these invisible regions and the metropole. Chapter 47's extended reflection on the novel form's capacity to facilitate social reform mobilizes the atmosphere to confront the spatial and epistemological problems of conceptualizing the entirety of a social body. The passage culminates with the invocation of the "good spirit" to remove the housetops and show how individual vice leads to the "raining [of] tremendous social retributions," but this is preceded by a complex consideration of the visibility and invisibility of social ills (702). While the "magistrate or judge" might "admonish the unnatural outcasts of society" from a distance, the narrator invites the reader to "follow the good clergyman or doctor" who has immediate experience of these "dens" of iniquity that lie "within the echoes of our carriage wheels" and to "breathe the polluted air, foul with every impurity that is poisonous to health and life" (700–1). This encounter with the physical spaces of depravity works to displace agency from the individual into the environment. "Vainly attempt," the narrator implores, "to think of any simple plant, or flower, or wholesome weed that, set in this foetid bed, could have its natural growth" (701). The following paragraph continues with this motif of the "polluted air" to further explore the causality of "unnatural" behavior and its spread. Drawing on the logic of miasma theory, the narrator suggests that while these spaces may seem isolated or cordoned off, they are nevertheless part of an interconnected web: "those who study the physical sciences . . . tell us that if the noxious particles that rise from vitiated air, were palpable to the sight, we should see them lowering in a dense black cloud above such haunts, and rolling slowly on to corrupt the better portions of town" (701). The hypothetical visibility embodied by the "if" is pushed one step further, so that the epidemiological is tied to the moral and "Vice and Fever" are seen to "propagate together": "how terrible the revelation" would be if "the moral pestilence that rises with [those clouds], and . . . is inseparable from them, could be made discernible too." The extended analogy structures a vision of systematic interconnection that is global in scope, as we might then see "how the same poisoned fountains that flow into our hospital and lazar-houses, inundate the jails, and make the convict-ships swim deep, and roll across the seas, and over-run vast continents with crime." The air becomes the flows of water that then wash over the lands.

The ecological logic of this powerful passage is worth examining in detail. On a basic level, the atmospheric imagery enables a movement outward from the local, uniting spaces that are both distinct and distant. Dickens deploys this same reasoning in an 1850 speech to the Metropolitan Sanitary Association to critique the belief that the local conditions that bred diseases such as cholera were the concern only of those within the immediate parish. After noting that contrasts in wealth and comfort were inevitable in "civilized communities" but not "afforded by our handsome streets, our railroads and our electric telegraphs, in the year of our Lord 1850," Dickens confronts the problem of thinking that fever and disease can be locally contained: "so long as [one] breathed the same air as the inhabitants of that court, or street, or parish,—so long as he lived on the same soil, was lighted by the same sun and moon, and fanned by the same winds, he should consider their health and sickness as most decidedly his business."[21] In both this address and *Dombey*, the air and atmosphere provide a way of conceptualizing interconnection in a manner that thinks beyond the arbitrary barriers presented by house-tops and parish boundaries. More complexly, the meteorological dimensions of this imagery also provide ways of understanding interconnection in the other direction—in how the local is conditioned by the distant. The emergence of meteorology in the mid–nineteenth century was made possible by the existence of an international telegraphy network that allowed observations and data to be shared quickly, establishing a relationship between particular local effects and the dynamics of the atmosphere. But as Katherine Anderson has shown, developing a science of the weather was not just about building a universal science out of particular and localized observations; it also required conceptualizing how a global system was responsible for producing intensely localized conditions. Anderson, for instance, cites an example of this increasing rigor in the 1848 meeting of the British Association for the Advancement of Science, where Edward Sabine presented the work of the German meteorologist Heinrich Dove, a major figure in nineteenth-century meteorology responsible, among other things, for compiling global temperature maps. Dove's weather model showed how European weather was produced by equatorial and polar air currents, which produced storms when they collided.[22] An 1851 piece in *Household Words* called "The Wind and the Rain," co-written by Dickens and Henry Morley, communicates similar theories, moving from immediate experience of a rainy day—"The wind to-day is blowing from the north-west, and it flings the rain against our window-panes. That boy, Tom, will be very wet, for he is out in it without an umbrella"—to an extended scientific discussion of how winds are generated by inequalities in temperature and, in turn, interact with rising water vapor to produce precipitation.[23] In short, the weather supports the reforming vision of *Dombey and Son* not simply through its rhetorical force (the "raining [of] tremendous social retributions") but through a logic of interconnection that lends this rhetoric its power.

However, if the weather provides the vehicle by which *Dombey and Son* conceptualizes a global system of interconnected locales, it also foregrounds the imaginative or fictive gesture required to situate the particular and immediate within a totalizing vision. The novel aims to rouse readers "to a knowledge of their own relation" to "the world of human life" but delimits the spatial terrain upon which such relations might be traced. This seeming contradiction lays bare the core problem of thinking of the novel in terms of a model or system, insofar as doing so entails abstraction and distance from the material particulars that are the principle component of realist representation. Put more simply, the work of the weather in the novel marks the limits of metonymy by revealing the problems that emerge at a scale as vast as the global. The same peculiar logic that Stewart argues governs the ideological work of empire in the novel—a logic "somewhere in the middle zone between metonymy and metaphor"—also structures the counter-ideology that incorporates this invisible region into an understanding of systemic interconnection. This logic is exhibited clearly, if still subtly, in the extended passage in Chapter 47, first through the collocation of "Vice and Fever" and then more explicitly through the elision of the difference between dark storm clouds and the metaphorical clouds of moral pestilence. If the "noxious particles" were visible, *then* "we should see them lowering in a dense black cloud," and if "moral pestilence" were visible, too, *then* we should see the "spread [of] contagion among the pure" (701).

The novel not only relies on this logic in its appeal to readers but also dramatizes it at two significant moments where characters read the sky and metonymic associations give way to a figurative and fictive sense of totality. Both occur as Walter Gay's friends ponder the uncertain fate of the *Son and Heir*. In Chapter 23, Florence, already grieving the death of her brother, hangs in an "agony of suspense" regarding Walter. "Uncertainty and danger seemed written upon everything," as the "weathercocks on spires and housetops were mysterious with hints of stormy wind, and pointed, like so many ghostly figures, out to dangerous seas, where fragments of great wrecks were drifting, perhaps" (357). As she enters the city, "pictures and prints of vessels fighting with the rolling waves filled her with alarm," and the "smoke and clouds . . . made her fear there was a tempest blowing at that moment on the ocean" (358). The nature of Florence's vision here wavers between the paranoid and the paratactic, displaying—with opposite effect—the same sensibility as her father's belief that "winds blew or against [his] enterprises" (12). The weathercocks point to the winds, which point to the seas where Walter's ship may be, but the link between the prints of the vessels and the *Son and Heir*, between the smoke and the tempest that may have led to its demise, exceeds metonymy. A similar scene occurs in Chapter 30 as Captain Cuttle paces the rooftop of the Wooden Midshipman "to take an observation of the weather," less uncertain about Walter's probable fate, but no less capable of mastering his fear. Although "the rain fell fast, and the wind blew hard," the Captain does not "associate the weather of that time with poor Walter's destiny." Yet he is nevertheless unable to master the feelings of despair as he scans the "wilderness of house-tops, and looked for something cheery there in vain" (496). If, for Florence, the weathercock initiates the train of associations that bridge the space between her and Walter, here the "crazy weathercock of a midshipman, with a telescope at his eyes, once visible from the street, but long bricked out" embodies the representational limits to metonymic systems grounded in house-tops and their removal.

As the sky becomes the vehicle for characters and readers alike to conceptualize the spatial dimensions of a global system, it also reveals the constitutive tension of a literary form that increasingly moves between concrete particulars and abstract systems, specific locales and the distant regions that produce them. The sky in *Dombey and Son* is not something, as Jesse Oak Taylor suggests of later Dickens and Victorian fiction generally, entirely of "our manufacture," nor is nature a realm entirely distinct from a coherently conceived social realm of individuals and institutions. Rather, the sky and the novel's other meteorological tropes move curiously between material and figurative registers, situating characters and events within wider contexts while marking the limitations of representing the full spatial dimensions and mediating structures of that system. The logic of this is presented most poignantly and tellingly by the name of the doomed *Son and Heir*, whose wreck figures the interrelation between the individual, the economic, and the ecological in the novel. The ship itself explicitly links individual agency and the processes of capitalism by conjoining Dombey's son, his firm, and his transactions with empire. The name, though, also conjures the homophonic "sun and air," which is both implicit but also displaced from these meanings. It points to the dominant means by which the novel works to recalibrate the relationship between the domestic and the economic while also registering an inescapable incommensurability of scale. While these junctures between concrete particulars and the systems in which they participate have been and continue to be the site where symptomatic reading finds a point of departure, we might also find in them new opportunities for exploring the affordances of the novel form to conceptualize structures of interrelation that can only be experienced indirectly, obliquely—through abstraction.

Notes

1. Allen MacDuffie, *Victorian Literature, Energy, and the Ecological Imagination* (Cambridge: Cambridge University Press, 2014), 89; see also, Jesse Oak Taylor, *The Sky of Our Manufacture: The London Fog in British Fiction from Dickens to Woolf* (Charlottesville: University of Virginia Press, 2016).

2. Suvendrini Perera, "Wholesale, Retail and for Exportation: Empire and the Family Business in *Dombey and Son*," *Victorian Studies* 33, no. 4 (1990),

605. See also, Jeff Nunokawa, "For Your Eyes Only: Private Property and the Oriental Body in *Dombey and Son*," in *Macropolitics of Nineteenth-Century Literature*, ed. Jonathan Arac and Harriet Ritvo (Philadelphia: University of Pennsylvania Press, 1991), 138–58.

3. Charles Dickens, *Dombey and Son*, ed. Andrew Sanders (New York: Penguin, 2002), 129. Hereafter cited parenthetically in the text.

4. Jason Moore, *Capitalism in the Web of Life: Ecology and the Accumulation of Capital* (London: Verso Books, 2015).

5. In thinking of systems and interconnection, I follow Bruno Latour's methodological emphasis on the tracing of connections between agents rather than the assumed, pre-given existence of the social or of systems. Political ecology, in his view, "*does not know* what does or does not constitute a system [and] does not know what is connected to what"; it is "incapable of integrating the entire set of its localized and particular actions into an overall hierarchical program, and it has never sought to do so." *The Politics of Nature*, trans. Catherine Porter (Cambridge: Harvard University Press, 2004), 21–22, original emphasis.

6. Charles Dickens, *Martin Chuzzlewit*, ed. Patricia Ingham (New York: Penguin, 1999), 5.

7. Working notes exist for both *The Old Curiosity Shop* (1840–41) and *Martin Chuzzlewit* (1843–44), but *Dombey and Son* is the first novel for which a complete set of number plans exist. For an introduction to Dickens's number plans for *Dombey and Son* (and the plans themselves), see Harry Stone, ed., *Dickens' Working Notes for His Novels* (Chicago: University of Chicago Press, 1987), 49–99.

8. Steven Marcus, *Dickens: From Pickwick to Dombey* (London: Chatto & Windus, 1965), 297–98. Garrett Stewart similarly suggests that "Never more openly than in *Dombey and Son* does the domestic melodrama of Dickensian fiction reach out to embrace an entire global vision." "The Foreign Offices of British Fiction," *Modern Language Quarterly* 61, no. 1 (2000): 203.

9. Elaine Freedgood, *The Ideas in Things: Fugitive Meaning in the Victorian Novel* (Chicago: University of Chicago Press, 2006), 90. My argument is indebted to the forms of metonymical reading Freedgood elaborates and, in particular, her extension of Bill Brown's analysis of the barometer Roland Barthes cites as an instance of the "reality effect" that draws out the capacity of such objects "to materialize . . . an absent presence" (11).

10. Audrey Jaffe, *Vanishing Points: Dickens, Narrative, and the Subject of Omniscience* (Los Angeles: University of California Press, 1991), 71.

11. Ayşe Çelikkol, *Romances of Free Trade: British Literature, Laissez-Faire, and the Global Nineteenth-Century* (Oxford: Oxford University Press, 2011), 4. See pages 123–42 for a reading of Dickens's *Little Dorrit* that focuses on the

way in which the novel "places global capitalism at the root of the dissolving distinction between the private and the public" (124).

12. Fredric Jameson, "Cognitive Mapping," reprinted in *Critical Theory: A Reader for Literary and Cultural Studies*, ed. Robert Dale Parker (Oxford: Oxford University Press, 2012), 469.

13. Stewart, "Foreign Offices," 186.

14. Ibid., 183–84.

15. Ibid., 204.

16. Mary Favret, "War in the Air," *Modern Language Quarterly* 65, no. (2004): 531–59. For meteorology and novelistic atmosphere, see also Justine Pizzo, "Atmospheric Exceptionalism in *Jane Eyre*: Charlotte Brontë's Weather Wisdom," *PMLA* 131, no. 1 (2016): 84–100, and Daniel Williams, "The Clouds and the Poor: Ruskin, Mayhew, Ecology," *Nineteenth-Century Contexts* 38, no. 5 (2016): 319–31.

17. A small provincial island of Portugal about 600 miles southwest of Lisbon, Madeira played a pivotal role in British foreign trade from the seventeenth century through the nineteenth. Britain dominated the Madeira wine trade, as bottles matured while serving as ballast on ships. See Desmond Gregory, *The Beneficent Usurpers: A History of the British in Madeira* (Cranbury, N.J.: Associated University Presses, 1988).

18. Stewart, "Foreign Offices," 203.

19. Moore, *Capitalism*, 36, emphasis original.

20. Michael Klotz, "*Dombey and Son* and the 'Parlour on Wheels,'" *Dickens Studies Annual* 40 (2009): 71.

21. K. J. Fielding, ed., *The Speeches of Charles Dickens* (Oxford: Oxford University Press, 1960), 106–7.

22. Katherine Anderson, *Predicting the Weather: Victorians and the Science of Meteorology* (Chicago: University of Chicago Press, 2005), 86–94; see also pages 1–14 and 171–233.

23. Charles Dickens, "The Wind and the Rain," in *The Uncollected Writings of Charles Dickens, Volume I*, ed. Harry Stone (Bloomington: Indiana University Press, 1968), 286. Morley, who had produced an article the prior week titled "The World of Water," wrote the large majority of this piece and seems primarily responsible for the more scientific elements of it.

PART III

Scale

CHAPTER 7

How We Might Live
Utopian Ecology in William Morris and Samuel Butler

Benjamin Morgan

At least since Friedrich Engels accused the socialist utopias of Joseph Fourier, Robert Owen, and Henri de Saint Simon of "drifting off into pure phantasies," utopian thought has had to defend itself against the charge that the worlds it builds are, at best, impossibilities and, at worst, imaginary distractions from real change.[1] Recuperating utopianism has been the project of some of the twentieth century's most influential thinkers, from Theodor Adorno's account of the aesthetic object as a "negative appearance of Utopia" in its rejection of the empirical world, to Fredric Jameson's claim that "the Utopian idea ... keeps alive the possibility of a world qualitatively distinct from this one and takes the form of a stubborn negation of all that is."[2] But while the majority of writing about utopia has tended to agree with Northrop Frye's observation that utopia "is primarily a vision of the orderly city and of a city-dominated society," we now find ourselves at a historical juncture where it is not ideas of justice, economics, and government, but the ecological systems they depend on that represent the most urgent object of utopian thought.[3] This is evidenced not only in geoengineering scenarios that would literally remake the world's climate (for instance by injecting massive amounts of heat-reflecting sulfur into the stratosphere) but also in countervailing calls for a reorganization of the global economy around the principle of "degrowth," an intentional "downscaling of production and consumption that increases human well-being and enhances ecological conditions."[4]

In view of these hopes that we might remake our economic and ecological worlds, it is worthwhile to revisit a Victorian utopian literary tradition that emerged partly in response to the destruction of Britain's atmosphere by coal-powered factories and of its landscape by the mines that provided their energy and railroads that moved their goods. In the coal era's literary history, no dyad of utopias is as significant as Edward Bellamy's *Looking Backward* (1888) and William Morris's *News from Nowhere* (1890): Bellamy's urban fantasy of centralized management inspired hundreds of "Nationalist Clubs," a journal, and a political party in America; while Morris's anarchic riposte is widely regarded as a foundational text of ecosocialism.[5] For Jameson, the binary of the "pastoral Morris, as opposed to the industrial Bellamy" offers ballast for the claim that "what uniquely characterizes this genre [utopia] is its explicit intertextuality: Few other literary forms have so brazenly affirmed themselves as argument and counterargument."[6] But the conventional Bellamy-Morris pairing may also obscure the extent to which Morris's work is representative of a broader field of nineteenth-century pastoral or idyllic utopianism, which torqued the urban political imagination of utopian thought with a new ecological concern brought on not only by industrialism but also by the post-Darwinian view that natural selection shaped human society.

In many regards, the ecological utopianism of the late nineteenth century can be understood as an aspect of a widely felt "urge to idealize a simple, rural environment" that Leo Marx identified as the pastoral response to industrialization in North America and Britain.[7] But if we align utopia's anti-industrial turn with the idealism of Eden, Cockaigne, and Arcadia, we risk falling back into Engel's notion of utopia as mere "phantasy"—despite the powerful critiques of the science–versus–utopianism opposition mounted by Jameson, Adorno, and others. This essay argues that the turn to nature in utopian literature around 1870 is better understood as an attempt to use structural features of utopian form to navigate the complexity of ecological and economic systems as they intersect at multiple scales. Utopian texts and projects grasp a relation among *closure, totality,* and *system*: By demarcating a threshold, they designate a self-contained world in which radically strange webs of exchange, association, and desire may be explored. This exploration is often marked by a shifting among scales: Literary utopias' emphasis on systems is capable of transitioning a reader's attention, for instance, between the micro-scale event of purchasing an item and macro-scale networks of labor and resource use.

Utopias may thus be understood as ecological not only insofar as they take up subject matter relating to transformed human-nature relations, but more abstractly in the sense that utopian form is intrinsically committed to depicting systemic relationality as such. Utopia translates into an aesthetic idiom modes of systems-thinking that the proto-ecologists Darwin and Marx described, respectively, as apprehending "the many complex contingencies, on which the existence of each species depends" or a "rich totality of many determinations and relations."[8] In order to highlight the ecological possibilities of utopian form, I turn away from the Morris-Bellamy pairing, instead reading *News from Nowhere* in relation to an earlier utopia that inspired it, Samuel Butler's *Erewhon* (1872). Butler's novel describes a lost society on a New Zealand–like island that had banned any technology more advanced than that used around the year 1600. Butler simultaneously satirizes the utopianism inherent to a mid-nineteenth-century settler discourse that imagined nature in terms of resources to be appropriated for private gain, and makes a powerful argument that distinctions between the natural and the social are chimerical in the face of Lamarckian and Darwinian evolutionary theory. Read with Morris, Butler's writing reveals that what is most significant about utopian responses to British industrialism and colonialism is not that these utopias offer up the merely palliative fantasy of an idyllic return to nature; nor—as Jameson and Darko Suvin have each argued of utopia and science fiction in general—that they provide the means for "apprehending the present as history" through the formal mechanism of "cognitive estrangement."[9] Rather, it is that utopia makes central the problem of representing the complex interactions between human and nonhuman systems by elucidating the ways in which nature pervades society at both micro and macro scales. Reading Butler and Morris together is important because the pairing resists a stark alternative of either a return to nature or a well-ordered city, foregrounding instead, through discourses of settler colonialism, evolutionary theory, and transnational socialism, the multiple scales at which human and nonhuman domains are intractably entangled. If utopia, as I will argue, is a form committed to the mediation of totality, Morris's and Butler's questioning of nature-society distinctions is not accidental or contingent to utopianism; instead, it is a strategy to expose any utopian totality conceived in exclusively social terms as unsatisfactory. I suggest in closing that this reading of nineteenth-century ecological utopianism as revealing the instability of nature/society dualism is valuable at present because it allows us to recognize a contemporary resurgence in environmental utopianism (for example, geo-engineering, rewilding, and degrowth) less as imaginary wish-fulfillment than as the animation of a specific affordance of utopia, as an aesthetic form, to mediate multisystemic complexity.

Utopian Systems

Critical accounts of the proliferation of utopian writing beginning in the 1870s have tended to focus on the economic and political conditions that made the idea of a radically reorganized society attractive to readers. Remarking on the coincidence that three major speculative texts were either published or submitted to publishers on May 1, 1871—Butler's *Erewhon*, Edward Bulwer-Lytton's *The Coming Race*, and George Chesney's *The Battle of Dorking*—Suvin argues that "there is no doubt that the immediate stimuli were the Franco-Prussian War and the Paris Commune of 1871, and in a more diffuse way the political regroupings in the UK attendant upon the 1867 suffrage reform."[10] Matthew Beaumont affirms Suvin's account, arguing further that late-Victorian "utopian thought is a product of the fact that revolutionary social change was, to all extents and purposes, impossible," and Kristin Ross has recently argued that Morris's *News from Nowhere* responds to the Commune by imagining that socialism would involve "communal luxury" rather than "the sharing of misery."[11]

Often left unremarked in these accounts is the consistency with which not only Morris but a wide array of utopian writing imagined that the outcome of political and economic impasses would be some form of a return to nature—a "return" that could be imagined either as a resolution or as a catastrophe. Richard Jefferies's *After London* (1886) is not only a story of Britain's imperial decline but also of the wild spaces that reassert themselves in Empire's absence. The gambit of H. G. Wells's more radical and far-reaching tale of what would come after London, *The Time Machine* (1895), was to doubly upend the trope of an idyllic return to nature by revealing the stupidly happy Eloi first to be degenerate humans and then to be food for Morlocks. W. H. Hudson's 1887 *A Crystal Age* (in which human society has taken the shape of a beehive) and Charlotte Perkins Gilman's 1915 *Herland* (a feminist utopia in which a parthenogenic, communal society of women have "improved their agriculture to the highest point, and carefully estimated the number of persons who could comfortably live on their square miles; [and] then limited their population to that number") each respond to Malthusian fears of overpopulation by imagining societies in which reproduction is tightly controlled to conform to natural boundaries.[12] The communal focus of the late-Victorian utopia can be understood as a response to prospective limitations upon growth imposed by the natural world or even to the idea of the eventual end of humanity in deep time. It was not only the political ideals of the Communards, but also William Stanley Jevons's warnings about peak coal and degenerationist discourse inspired by Darwin and Spencer that furnished the imaginary infrastructure of late-nineteenth-century revival of literary utopias.

Recent scholarship has advocated for a return to the ideas of two of the most significant nineteenth-century ecoutopians, John Ruskin and William Morris, either as modeling an early attempt "to exit the fossil fuel economy and consumer society" (Vicky Albritton and Fredrik Albritton Jonsson) or as beginning a "prescient" discussion of "fossil fuels' effects on the atmosphere" (Elizabeth Miller).[13] Part of what makes Morris and Ruskin attractive today is their commitment to the idea that the economy or the human-nature relation might be radically reconfigured. This eco-socialist utopianism is especially apparent in "How We Live and How We Might Live," one of a number of lectures Morris gave in the 1880s and then published either in *Commonweal* or as pamphlets. Implicitly adopting a utopian frame, the lecture takes flight from the idea that those interested in socialism may reasonably demand "at least some idea of what that life may be like" after the end of capitalism.[14] In effect, what Morris describes in the essay is a broken global system in which the industrial exploitation of nature also exploits populations in the East and Global South:

> The Indian or Javanese craftsman may no longer ply his craft leisurely. . . . in producing a maze of strange beauty on a piece of cloth: a steam-engine is set a-going at Manchester, and that victory over Nature . . . is used for the base work of producing a sort of plaster of china-clay . . . and the Asiatic worker, if he is not starved to death outright, as plentifully happens, is driven himself into a factory to lower the wages of his Manchester brother worker.[15]

What is especially notable here is that Morris understands the harnessing of coal energy to fuel textile factories as having effects that extend far beyond Manchester's pollution or the mined English landscape: His view of ecology and economy is in this instance global and systemic. Hence a recognition that technological fixes cannot in themselves provide social fixes: "The conquest of Nature is complete . . . and now our business is . . . the organization of man, who wields the forces of Nature."[16] Morris is clear that it is not the steam engines that are the problem, but the way in which they mediate human relations: "it is the allowing machines to be our masters and not our servants that so injures the beauty of life nowadays."[17] Morris then imagines that in a transformed socialist society—that is, the utopian future in which we "might live"—machinery will be relied upon insofar as it obviates the need to do unpleasant work (for example, coal-mining), and that at some phase "handwork rather than machinery" may be preferred for certain tasks that afford pleasures of craft.[18]

What Morris is therefore aspiring toward, as Raymond Williams argues in *Culture and Society*, is not reducible to the equation "Morris—handicrafts—get rid of the machines."[19] For Morris, "machinery" does not name a category of industrial objects; rather, it is the total system of relations connecting humans, industrial technology, and nonhuman nature—a shorthand term for "our control of the powers of Nature."[20] To emphasize Morris's embrace of some form of machinery, Williams goes so far as to imaginatively censor the most famous works within Morris's oeuvre: "I would willingly lose *The Dream of John Ball* and the romantic socialist songs and even *News from Nowhere*—in all of which the weaknesses of Morris's general poetry are active and disabling" if this meant more attention would redound to essays like "How We Might Live."[21] While this claim drew objections from E. P. Thompson, it is illuminating for revealing the difficulty of recognizing that Morris's romanticism allowed for the technological conquest of nature.[22]

Morris's turn to utopianism in order to understand large-scale systems highlights closure or boundedness as a formal affordance of utopia more generally. Morris's strategy closely aligns with Roland Barthes's comments on the similarity between the Marquis de Sade's cloistered pleasure retreats and Fourier's phalansteries. Barthes argues that for the utopian social form, "the enclosure permits the system": That is, for both Sade's retreats and Fourier's phalansteries, utopian thought can conceive of systems of actors, practices, and pleasures only by demarcating an autonomous space within which they circulate.[23] This act of closure, emblematized by King Utopus's digging of a trench to create the first utopian island, makes the "threshold" a privileged figure for literary and nonliterary utopias alike: the boundary defines the system as a whole, making its totality an available object of representation. Barthes then aligns utopianism with the everyday in a way that implies that the programmatic or abstract elements of utopian writing acquire force only as specific transformations of the smallest lived practices. Echoing Ernst Bloch's identification of a utopian impulse in everyday acts of self-transformation (makeup, aspirin, clothing, and so on), Barthes argues that Utopia "is measured far less against theoretical statements than against the organization of daily life, for the mark of utopia is the everyday; or even: everything everyday is utopian: timetables, dietary programs, plans for clothing, the installation of furnishings, precepts of conversation or communication."[24] This idea of totality as produced by closure becomes central to Jameson's thought about utopia as negating actually existing systems (often capitalism) through this act of closure. For Jameson, "totality is then precisely this combination of closure and system" and "utopian form is itself a representational meditation on . . . the systemic nature of the social totality."[25] "System" here names the thought of a world in which a change to any part creates unpredictable changes to the whole: "the world is one immense and self-sufficient system—change anything in it, no matter how small, and the rest will necessarily be altered in unexpected ways."[26] Hence for Jameson and Barthes alike, two intersecting features at widely divergent levels: totality and everydayness; the system as a whole and the lived experience of it.

There is of course a significant disjuncture between "furnishings" or "dietary programs" and an all-encompassing "totality" or "theoretical statements." It is useful to rephrase Barthes's and Jameson's claims about the totality-system nexus and the "everydayness" of utopian thought as a scalar shift that utopian writing and utopian thought must continually carry out, between highly localized ethnographic thick descriptions of objects and practices, and accounts of the functional relations of these objects and practices within the totality of a utopian world-system. We see such a shift when Morris, in "How We Might Live," places the Manchester factory in relation to an Indian weaver: Morris's utopian imagination is premised upon this apprehension of systemic interconnectedness. It is in this sense that *system* and *scale* intersect: utopian writing works by showing how a large-scale social and ecological system is constituted by the manifold practices that occur in local and specific situations.

This practice of systemic scale shifting is a structuring feature of *News from Nowhere*, which repeatedly oscillates between ethnographic descriptions of Nowhereian society and accounts of the social totality. In a chapter titled "A Little Shopping," William Guest is given a pipe "carved out of some hard wood very elaborately and mounted in gold sprinkled with gems," which he exclaims is "altogether too grand" and worries about losing it.[27] To this, the shopkeeper responds that should he lose it, "somebody is sure to find it, and he will use it, and you can get another"; and then reacts with confusion at Guest's expostulation that he has no money to pay for it.[28] A common object, the pipe not only figures everydayness, but, more specifically, the utopian transformation of the everyday (into "gold sprinkled with gems") by a system in which private property and money are alike abolished, and in which production is driven by the pleasure of craft rather than monetary profit.[29] In this regard, Morris's pipe is functionally the opposite of Flaubert's barometer as read by Barthes, not signifying "the category of 'the real'" through its semantic emptiness, but rather opening up the unrealized possibility of a transformed world-system through its semantic density.[30] The pipe's sensuous materiality is freighted with the task of mediating the two scalar registers of utopian writing, at once offering Guest the immediate sensory pleasures available within a rationally reorganized economic system (micro) and revealing the dynamics of that system as a whole (macro). Here and elsewhere, the genre of utopia acquires force not only, as Jameson argues, for its capacity to defamiliarize and historicize the present or to formally negate the given (regardless of the actual feasibility of utopia's content), but also for its capacity to depict scalar shifts within a given world-system.

Combined with the attention to world-scale systems reflected by "How We Might Live," the scale shifting in relation to the pipe gives us a new way of understanding why the figure of the garden should be so central to Morris's utopia. Guided by a notion of machinery as a means of mediating human relations via the mastery of nonhuman nature, Morris moves beyond the idea of utopian totality as purely social (Frye's "city-dominated society") by making natural systems an explicit problem for utopian thought.[31] Although *News from Nowhere*, whose characters frequently express a "passionate love of the earth," is often thought of as idealizing a return to nature, in fact the version of "nature" operative within the novel is so universally cultivated and controlled as to decisively undermine the opposition natural/artificial.[32] Recall several features of Morris's bucolic England that may pass without notice: It is at least as populous as that of the nineteenth century; signs of human civilization are so pervasive that "it is not easy to be out of sight of a house"; even the "wild nature" that does remain in this future England is largely treated as a resource for "timber" and "grazing"; and as Guest travels up the river, "all along the Thames there were abundance of mills used for various purposes."[33] Cultivation is so widespread that, strangely, wild nature may be experienced as the simulacrum of a designed landscape. Hammond, one of Guest's guides, observes: "I have heard that they used to have shrubberies and rockeries in gardens once; and though I might not like the artificial ones, I assure you that some of the natural rockeries of our gardens are worth seeing."[34]

Why this entirely cultivated garden rather than the wilderness imagined in Jefferies's *After London*, a novel that Morris read with "absurd hopes," and that according to an early biographer, Morris thought "represented very closely what might really happen in a dispeopled England"?[35] The figure of the garden matters to Morris not as a romanticized image of a world of Edenic abundance where all desires are providentially satisfied by generous Nature, but as humanity's total and harmonious control of natural systems. This means that processes within the natural world, rationally dominated and mastered in the very way suggested by "How We Live," take the place of "machinery" within nineteenth-century industrial society. The "delicious super-abundance of small well-tended gardens" near the end of the journey up the Thames, the "orchard . . . of apricot trees" in Trafalgar square: These are not the opposite of but figures for machinery—that is, nature's machinery deployed without entailing intra-human exploitation.[36] From this perspective, Suvin's claim that Morris "overreacted into a total refusal to envisage any machinery, technological or societal. . . . England is now a garden" rests on a distinction between machine and garden that is not supported by the text itself: The garden is, in a very meaningful way, a new, humane form of machinery.[37]

But as distinct from Guest's pipe, which is a synecdoche of the transformed economic system, the garden and the orchard are simultaneously the part and the whole; like a fractal, to zoom out from the orchard in Trafalgar square is only to see, again, the orchard that is England. Nothing lies beyond the garden because the garden is the new totality. It is only punctured by Guest himself, whose conspicuous name already signals that he will be treated as though he is from "some distant country" or "another planet" and whose true origin is with rare exception treated as unspeakable.[38] Nowhere's garden is troubled by and ultimately rejects his alien presence, and through this rejection reveals a claustrophobic anxiety about foreignness as such. To fully understand the effects of this closure to a world beyond requires looking backwards to a text that shared Morris's interest in a utopian relation between social and natural systems but that scaled up its world beyond the walled garden to include Britain's colonial possessions.

Beyond the Garden

Erewhon, along with More's *Utopia* and Jefferies's *After London*, was one of the primary sources for *News from Nowhere*; Morris read Butler's novel aloud with Philip Webb and Edward Burne-Jones in autumn of 1882, and May Morris recalls that in her youth "Butler's *Erewhon* was a household word."[39] Butler's utopia, like Morris's, is nominally medievalist, in that Erewhonians have banned modern technology such as steam engines and even watches. But by contrast with Morris, Butler's "nature" is not that of the English countryside but of New Zealand settler colonialism: a resource to be appropriated rather than a landscape to be enjoyed. Butler thus makes explicit the extent to which utopianism may be captured by an imperialist logic of extracting natural resources and human labor from distant colonies, even as his engagement with Darwinian thought exposes society itself as an extension of biological process and thus indistinct from nature. Through these means, Butler rewrites the boundaries that enclose the system, expanding and questioning in advance the image of nature that inheres in Morris's garden.

One explanation for this divergence lies in the type of political utopianism that matters to Butler: not utopian socialism, but rather the utopianism of mid-nineteenth-century New Zealand settler discourse. Sue Zemka has insightfully described this as a "myth of idyllic expansion": the notion that "the earth harbored large tracts of fertile and unpopulated land that were fit destinations for the surplus populations of Britain and Europe."[40] Initiated by Edward Gibbon Wakefield's promotion of New Zealand's colonization, it was taken up in a wide array of writing that explicitly deployed utopian tropes. Thomas Cholmondeley's *Ultima Thule; Or, Thoughts Suggested by a Residence in New Zealand* (1854) promised "a fresh life sent from Heaven"; the Owenite Robert Pemberton's pamphlet *The Happy Colony* (1854) proposed New Zealand as the "spot for the first stone of the temple of happiness to be laid," where a classless society might eventually allow students to spend the early decades of their lives learning eight languages in "Elysian Academies."[41] Other books and pamphlets were more practical but arguably no less utopian: William Stones's *New Zealand (The Land of Promise) and its Resources* (1863) described the island as "offering to industry easy competence, gradual improvement, and eventual affluence," while also offering detailed practical advice to prospective emigrants about natural resources, husbandry, and land regulations.[42]

It was easy for Europeans to think of New Zealand as a utopian space due to entrenched literary tropes that located paradises at the extreme northern or southern regions of the earth (*Ultima Thule*, or furthest north); Bloch explains the antipodal utopia as deriving from the idea that "the seat of the source of life in general was presumed to lie in the south, which knows the early spring earlier and from which the summer approaches."[43] This literary tradition partly accounts for why it was the case, in the words of the historian James Belich, that New Zealand "booster literature had a paradise complex. It portrayed new lands as biblical Lands of Canaan, Lands of Goshen, and Gardens of Eden."[44] The cultural significance of New Zealand as a utopian site can be measured both by the fact that *Looking Backward* appears to have borrowed its frame narrative from a little-known anonymous New Zealand utopia, *The Great Romance* (1881), and that New Zealand is the inspiration for Anthony Trollope's satirical utopia *The Fixed Period* (1882), which notably tempers its account of colonial political innovation by entertaining the possibility of forced euthanasia for citizens who reach the age of 67.[45]

By contrast with Morris's communal eco-socialist garden in which the successful management of nature has made possible the abolition of private property, New Zealand utopianism was premised on the idea that natural systems were available for appropriation in the service of accumulating capital. Butler's own autobiographical account of New Zealand settler life, *A First Year in Canterbury Settlement* (1863), a series of letters published by his father purportedly against Butler's wishes, frequently exhibits this point of view: Butler concludes with a second-person address to the reader that offers advice about where to locate a homestead, explains how to herd and shear sheep, and concludes with the promise that "if you have tolerably good fortune, in a very short time you will be a rich man."[46] Throughout his letters, Butler frequently shifts between a close attention to the financial and practical realities of settler life in New Zealand and evocative descriptions of the region's sublime scenery. The early letters, mulling over what to do with the money that Butler's father had finally given him after years of feuding, carefully tabulate the expected return on various investments: Putting five hundred sheep out on terms over seven years will yield £1,067 on an initial investment of £625; the higher-risk strategy of investing in a sheep run might eventually yield £2,000 per year on an initial investment of £6,000 (investing in land to establish an estate is judged too risky).[47] Butler's accounting is interspersed with powerful nature writing about his explorations of the South Island, including, notably, an extended description of his first view of Mount Cook as "a massy parallelogram, disclosed from top to bottom in the cloudless sky, far above all the others. . . . Mount Blanc himself is not so grand in shape."[48] But then, in a passage that hints at the irony that would be developed in *Erewhon*, Butler apologizes to the reader: "I am forgetting myself into admiring a mountain which is of no use for sheep. This is wrong. A mountain here is only beautiful if it has good grass on it. Scenery is not scenery—it is 'country,' *subauditâ voce* 'sheep.'"[49] Settler utopianism involves a view of nature in which ownership of part of a natural system (here, the transformation of sheep and pasture into more sheep) yields resources such as wool that can be metabolized into capital—and it was indeed Butler's success in appropriating nature that gave him the income necessary for pursuing his desire to write and paint. Like the utopianism of *News from Nowhere*, the utopianism of the New Zealand settler requires a detailed understanding of how natural systems and economic systems are enfolded into one another—not, however, in order to critique the industrial or individualist capture of nature, but instead to more successfully accumulate capital.

Erewhon satirizes New Zealand utopianism via a narrator who advocates capturing the people he discovers in a previously unknown civilization (whom he maintains to be the lost ten tribes of Israel) and placing them in the forced service of "religious sugar growers"—so as to both enrich himself and obtain "an immortal crown of glory" as the messianic restorer of the lost tribes.[50] But by contrast with *News from Nowhere*, in which the universal garden is the enabling mechanism of the utopian world, ideas of nature in *Erewhon* may seem peripheral to the utopia itself, in that the type of locodescriptive nature writing that appears throughout *First Year* is almost entirely relegated to a frame narrative in which Higgs (we learn the narrator's name in the sequel, *Erewhon Revisited* [1901]) battles the landscape and the elements on his way to Erewhon. Higgs spends most of his time in what seems to be—despite the ban on modern technology—a more or less urban environment. Here he chronicles aspects of a world that is, in keeping with the tradition of urban utopianism Frye discusses, social before it is natural: Readers learn about systems of justice, medicine, birth and death rituals, banking, religion, morality, education, technology, and, in chapters added to a revised edition, animal and vegetable rights. Furthermore, by contrast with William Guest's entrance into Nowhere by way of a marvelously unpolluted Thames and beautiful summer day, Higgs's first view of Erewhon aligns with the urban utopianism later most famously expressed in Julian West's view of a future Boston in *Looking Backward*. Higgs escapes from the dangers of steep mountains and wild rivers to witness first a pastoral scene of shepherds and then a built landscape: "fading away therein were plains, on which I could see many a town and city, with buildings that had lofty steeples and rounded domes."[51]

But to interpret Butler's novel with *Looking Backward* as an urban predecessor of Morris's pastoral would be to overlook the more fundamental argument Butler makes against a dualism of nature and society. Many of the judicial, economic, and political systems described in Erewhon are sharply satirized precisely for failing to recognize that society is an extension of nature rather than clearly distinct from it. Two aspects of Erewhonian society most clearly reveal the extent to which natural processes pervade human systems: first, an inversion of crime and illness such that crime is understood as a treatable malady beyond one's control but the sick are held legally responsible and imprisoned for their illnesses; and, second, the speculation that machines may someday evolve intelligence superior to that of humans and must therefore be severely limited.[52]

Alcoholism provides Butler with an illuminating limit case for the first point. The character Mahaina hides her crime of chronic stomach illness by pretending to be an alcoholic. She thus deploys an ambiguity specific to alcoholism: Does it reflect an individual moral weakness or a biologically inherited predisposition? Her subterfuge inspires Higgs to expose—ironically and unwittingly—the equally backward logics of both Erewhonian and British approaches to health and crime:

> Was there nothing which I could say to make them feel that the constitution of a person's body was a thing over which he or she had at any rate no initial control whatever, while the mind was a perfectly different thing, and capable of being created anew and directed according to the pleasure of its possessor? Could I never bring them to see that while habits of mind and character were entirely independent of initial mental force and early education, the body was so much a creature of parentage and circumstances, that no punishment for ill-health should be ever tolerated. . . ?[53]

These questions exactly invert the view that Butler espoused in his later writings on evolution, which adhered increasingly to a Lamarckian account of the biological transmissibility of habit. For Butler, to speak of habits of mind and character as "entirely independent of initial mental force and early education" is an absurdity; quite the contrary, he proposes in *Life and Habit* (1878), mind and character are already largely determined when a person is an embryo: "it would seem probable that all our mental powers must go through a quasi-embryological condition . . . and that all the qualities of human thought and character are to be found in the embryo."[54] It is upon the premise that individual character originates in the fertilized embryo (or even before) that any distinction between crime and illness becomes untenable: One has no more power over physical deficiencies than characterological ones.

This ironic inversion of illness and crime intends to reveal that "nature" is neither the picturesque scenery of idyll and pastoral nor the economic resource of settler utopianism—but is instead biologically woven through the very fabric of society. There is no such thing as a purely "social" phenomenon; all thoughts, habits, and practices are to some extent traceable to biological origins. It is from here that we may begin to see how the utopian form of *Erewhon* allows for a more nuanced ecological literacy than the frame narrative might initially imply. Despite their markedly different political and philosophical commitments, Morris and Butler each use the utopian form to engage in practices of scale-shifting within inextricably intertwined social and natural systems so as to expose dominant ideas of autonomous individuality as chimerical. But where for Morris, this thought takes the shape of an objection against the alliance of self and property, for Butler, it involves the use of Lamarckian evolutionary theory to expose individual actions as extensions of ancestral habits that precede the moment of birth. The systemic and ethnographic focus of utopian writing allows this overtaking of agency by heredity to become a direct object of representation for the novel because utopia prioritizes discussions of large-scale systems over the representation of psychological interiority. Utopianism thus matters not for the content of its promise, but because of its formal capacity to make visible both the arbitrariness and the wide-ranging effects of where boundaries are drawn between society and nature.

The best-known passages of *Erewhon* make clear that this examination of the imbrication of natural and social systems depends upon imaginary scale-shifting between local instances and temporally and socially wide-reaching systems. In three chapters titled "The Book of the Machines," originally published as articles in the Christchurch *Press* after Butler had read *On the Origin of Species*, Butler develops a philosophically prescient application of evolutionary theory to technology. These chapters explain the medieval state of Erewhonian technology by recounting the argument against machinery that had occasioned a revolution five hundred years prior to Higgs's arrival. It is argued that the line demarcating inanimate, unconscious machinery from conscious animal life is false; what we refer to as emotional and intellectual life is properly understood as "molecular action" and so, by extension, gradual modifications of machinery may lead to "the descent of conscious (and more than conscious) machines from those which now exist."[55] Coupled with the inevitable evolution of machine consciousness, the ever-increasing dependence of humanity upon technology justifies the elimination of all machinery corresponding to that invented later than the European sixteenth century. Otherwise, the evolution of machine intelligence means that humans will be "gradually superseded by our own creatures, till we rank no higher in comparison with them, than the beasts of the field with ourselves."[56]

It is striking that the reversion to a medieval society motivated by a fear of mastery by machines almost exactly corresponds to the historical change that Morris would later envision in "How We Live" and *News from Nowhere*—with the difference that Butler derives his reasoning from Darwinian thought and Morris from Marx and Ruskin. In each instance, the utopian form affords a scaling up of historical time to and beyond the Braudelian *longue durée* in order to consider the erosion of a distinction between humanity and nature. Utopia here is less a Blochian wish for improvement or transformation (indeed, the future imagined in "The Book of the Machines" must be avoided at all costs) than a mechanism for representing the large-scale or long-term effects of an artifact as innocuous as a watch. To align Morris and Butler is to recognize that neither is willing to reduce the concept of nature to pleasing landscape, which is a site of cultivation for Morris and a site of extraction capitalism for Butler. But, through *Erewhon*'s ambivalence toward settler colonialism, Butler pushes even harder against a utopian impulse that would find its fulfillment in nature's abundance: The protagonist who goes looking for an unclaimed pasture ironically finds instead a civilization—a civilization whose narrative function is, in turn, to expose the instability of the distinction between nature and society.

These dynamics speak to the fact that the turn to nature in late-nineteenth-century utopianism was not a cultural expression of imaginary wish fulfillment but a type of ecological thought that depicted intersecting human-nonhuman systems and shifts of scale within them. Indeed, it may seem that "pastoral" is not an adequate term either for Morris's garden or Butler's evolutionary and hereditary theory. To be sure, Morris's elucidation of socialist utopianism via the rustic simplicity of Nowherians fits neatly with William Empson's conception of pastoral as the routing of complexity through simplicity.[57] But deeper scientific and political literacies stemming from Darwin and Marx are at work for both writers in ways that make nature something more real than an image or a trope. Rather than romanticizing nature as a "simple vision of natural plenty" or a "natural delight in the fertility of the earth"—as Raymond Williams glosses classical pastoral—Morris and Butler examine and interrogate as arbitrary or constructed received distinctions between the social and the natural— either through the putative machinery of the garden or the evolutionary likelihood of machine intelligence and the false distinction between illness and crime.[58] Indeed, *Erewhon* explicitly mocks the pastoral ideal by concluding with a plan by Higgs (originally a shepherd, after all) to forcefully conscript Erewhonians to work at low wages on sugar plantations run by Christians, an act of violent exploitation justified as a means of caring for their souls: an insidious, scaled-up version of the pastoral care in which unfortunate Erewhonians take the place of sheep. Reading Butler with Morris suggests more generally that a significant contribution of late-nineteenth-century ecological utopias was to extend the utopian form beyond its concern with what Jameson describes as the "systemic nature of the social totality" to examine the ways in which social totality could no longer exclude nature in a modernity that was inflected by post-Darwinian thought and the colonialist transformation of natural resources into capital.[59]

Remaking the World

Drawing attention to the history of making nature an object of utopian speculation is useful in relation to ecological thought today because it allows us to make a distinction between the content of utopian projects, and utopianism as an analytic tool whose vehicle is often aesthetic form. At a moment of widespread crisis in connected ecological systems—not only global warming, but also the related processes of ocean acidification, biodiversity collapse, and nitrogen cycle disruption—utopianism has become highly controversial. The controversy derives in part from the widely critiqued position of "ecopragmatists" or "ecomodernists," such as Michael Shellenberger and Ted Nordhaus, whose Breakthrough Institute places faith in "technology and modernization" while espousing "a positive, optimistic paradigm called ecomodernism, which embraces modernity to leave more room for nature and expand human prosperity." A similar position is espoused by the ecologist Erle Ellis and his colleagues, whose *Ecomodernist Manifesto* proclaims the "conviction that knowledge and technology, applied with wisdom, might allow for a good, or even great, Anthropocene."[60] It is not only self-proclaimed ecomodernists who engage in a new utopianism; recent environmental writing has witnessed a shift from scenarios of gloom to scenarios of hope. E. O. Wilson's proposal of "committing half of the planet's surface to nature" (a massive increase from current nature preserves that cover 15 percent of the Earth's area and 2.8 percent of its ocean area) at once reflects a pastoral aspiration to re-wild much of the earth and embraces a form of hope that can only be described as utopian in the face of current geopolitical realities.[61] As always, utopianism elicits skepticism. For Clive Hamilton, faith in the technological fixes of geoengineering can only be seen as symptomatic of a delusional optimism that risks becoming complicit with right-wing denialism insofar as it gives up on moderate, real, and painful fixes that would require immediate change in favor of distant, uncertain, and risky technological solutions that require of most people no action today.⁶² In certain regards, this debate echoes Engels's distinction between utopian and scientific socialism: contemporary ecoutopianism comes under critique because, far from creating harmless castles in the clouds, it throws up real barriers to fomenting change.

Indeed, the resonances between current and past utopianisms are so evident that one might even expect that questions about ecological utopianism would have arisen more explicitly in the many recent Marxist critiques of the "Eurocentric" Anthropocene concept as reliant "on well-worn notions of resource- and technological-determinism"—critiques that have instead proposed that the proper designation for the present geological era is "*Capitalocene*, the historical era shaped by relations privileging the endless accumulation of capital."⁶³ When utopianism appears in this context, it is often implicitly, as an accusation to be avoided. Andreas Malm concludes his *Fossil Capital* with an argument in favor of a planned transition to an energy economy based upon the abundant power of wind and sun: "the fuel is already there, free for the taking, a 'gift of nature,' . . . to speak with Marx."⁶⁴ Malm is careful to argue that such a transition is more realistic than alternatives: eco-Marxists, Malm argues, must look beyond the easy slogan "'one solution—revolution,'" since "any proposal to build [socialism] on a world scale before 2020 and *then* start cutting emissions would be not only laughable, but reckless," and Malm asserts that the various geoengineering scenarios are riskier and more utopian than the notion of a planned economy.⁶⁵

But perhaps what Morris and Butler's version of utopian writing as effecting scalar shifts within complex systems teaches us is that utopianism is best thought of not as an idle optimistic affect but as a robust aesthetic strategy for the representation of long-term effects of complex systems. Following Jason Moore, adherents of the Capitalocene concept often work at the intersection of Immanuel Wallerstein's world-systems analysis and green Marxism in order to understand nature not as a resource that capitalism exploits but rather as a matrix within which capitalism emerges. From this perspective, one can redescribe ecological destruction in terms not just of "world-systems" but of "world-ecologies co-generated by each phase in the history of the world-economy. These equally show that the prosperity of the rich countries is constructed by way of a monopolization of the benefits of the Earth and an externalization of environmental damages."⁶⁶ This dynamic was, in effect, what Butler satirized by making the colonialist valences of pastoral utopianism a central problem in *Erewhon*.

Against the association of utopia with idyllic world-building, then, I would argue that "utopia" names the literary and aesthetic mode in which what Moore describes as world-ecology becomes available to representation. This means that it may be possible to evaluate speculative and utopian thinking, past and present, not for the viability of the solutions it proposes or for its capacity to induce an experience of historical consciousness, but as a literary and aesthetic means for representing complex, interlocking systems that must be conceived at multiple scales simultaneously. Utopia matters today not primarily as the site of the impossible, but as one of the only aesthetic forms capable of mediating totality.

Notes

1. Friedrich Engels, "Socialism: Utopian and Scientific," in *Karl Marx and Frederick Engels: Collected Works*, vol. 24, ed. Edward Aveling (New York: International Publishers, 1975), 290.

2. Theodor W. Adorno, *Aesthetic Theory*, ed. Gretel Adorno and Rolf Tiedemann, trans. Robert Hullot-Kentor (Minneapolis: University of Minnesota Press, 1997), 130; Fredric Jameson, *Marxism and Form: Twentieth-Century Dialectical Theories of Literature* (Princeton: Princeton University Press, 1971), 111.

3. Northrop Frye, "Varieties of Literary Utopias," *Daedalus* 94, no. 2 (Spring 1965): 325.

4. Paul J. Crutzen, "Albedo Enhancement by Stratospheric Sulfur Injections: A Contribution to Resolve a Policy Dilemma?," *Climatic Change* 77, nos. 3–4 (July 25, 2006): 211; François Schneider, Giorgos Kallis, and Joan Martinez-Alier, "Crisis or Opportunity? Economic Degrowth for Social Equity and Ecological Sustainability. Introduction to This Special Issue," *Journal of Cleaner Production* 18, no. 6 (April 2010): 512.

5. On Morris's anticipation of "socialist ecologies," see Florence Boos, "An Aesthetic Ecocommunist: Morris the Red and Morris the Green," in *William Morris: Centenary Essays*, ed. Peter Faulkner and Peter Preston (Exeter: Exeter University Press, 1999), 22.

6. Fredric Jameson, *Archaeologies of the Future: The Desire Called Utopia and Other Science Fictions* (New York: Verso, 2005), 159, 2.

7. Leo Marx, *The Machine in the Garden: Technology and the Pastoral Ideal in America* (New York: Oxford University Press, 2000), 5.

8. Charles Darwin, *On the Origin of Species*, ed. Gillian Beer (New York: Oxford University Press, 2008), 237; Karl Marx, *Grundrisse: Foundations of the Critique of Political Economy*, trans. Martin Nicolaus (London: Penguin, 1993), 100.

9. Jameson, *Archaeologies*, 288; Darko Suvin, *Metamorphoses of Science Fiction: On the Poetics and History of a Literary Genre* (New Haven: Yale University Press, 1979), 4.

10. Darko Suvin, "Victorian Science Fiction, 1871–85: The Rise of the Alternative History Sub-Genre," *Science Fiction Studies* 10, no. 2 (1983): 148.

11. Matthew Beaumont, *Utopia Ltd.: Ideologies of Social Dreaming in England, 1870–1900* (Leiden: Brill, 2005), 28; Kristin Ross, *Communal Luxury: The Political Imaginary of the Paris Commune* (London: Verso, 2015), 65.

12. Charlotte Perkins Gilman, *Herland and Related Writings*, ed. Beth Sutton-Ramspeck (Peterborough: Broadview Press, 2013), 104.

13. Vicky Albritton and Fredrik Albritton Jonsson, *Green Victorians: The Simple Life in John Ruskin's Lake District* (Chicago: University of Chicago Press, 2016), 12; Elizabeth Carolyn Miller, "William Morris, Extraction Capitalism, and the Aesthetics of Surface," *Victorian Studies* 57, no. 3 (2015): 398, 399.

14. William Morris, "How We Live and How We Might Live," in *The Collected Works of William Morris: Signs of Change and Lectures on Socialism*, vol. 23 (London: Longmans, Green, 1915), 4.

15. Ibid., 8.

16. Ibid., 15.

17. Ibid., 24.

18. Ibid.

19. Raymond Williams, *Culture and Society, 1780–1950* (New York: Anchor Books, 1960), 167.

20. Morris, "How We Live," 24.

21. Williams, *Culture and Society*, 167–68.

22. See Edward Thompson, "Romanticism, Utopianism and Moralism: The Case of William Morris," *New Left Review*, no. 99 (1976): 99.

23. Roland Barthes, *Sade, Fourier, Loyola* (Berkeley: University of California Press, 1989), 17. On utopian self-transformation, see Ernst Bloch, *The Principle of Hope*, vol. 1 (Cambridge: MIT Press, 1995), 345.

24. Barthes, *Sade, Fourier, Loyola*, 17.

25. Jameson, *Archaeologies of the Future*, 5, xii.

26. Ibid., 77.

27. William Morris, *News from Nowhere, Or, An Epoch of Rest: Being Some Chapters from a Utopian Romance*, ed. David Leopold (Oxford: Oxford University Press, 2003), 32.

28. Ibid.

29. Ibid.

30. Roland Barthes, *The Rustle of Language* (New York: Hill and Wang, 1986), 148.

31. Frye, "Varieties of Literary Utopias," 325.
32. Morris, *News from Nowhere*, 178.
33. Ibid., 63, 64, 168.
34. Ibid., 64.
35. William Morris, "To Georgiana Burne-Jones, April 28, 1885," in *The Collected Letters of William Morris*, vol. 2, ed. Norman Kelvin (Princeton: Princeton University Press, 1987), 426; John William Mackail, *The Life of William Morris*, vol. 2 (London: Longmans, Green, 1901), 144.
36. Morris, *News from Nowhere*, 173, 36.
37. Suvin, *Metamorphoses of Science Fiction*, 181.
38. Morris, *News from Nowhere*, 18, 47.
39. Mackail, *Life of William Morris*, 90; William Morris, *The Collected Works of William Morris*, vol. 22 (London: Longmans, Green, 1915), xxvii. On Morris's "warm" response to Butler, see also E. P. Thompson, *William Morris: Romantic to Revolutionary* (New York: Pantheon Books, 1977), 802.
40. Sue Zemka, "*Erewhon* and the End of Utopian Humanism," *ELH* 69, no. 2 (2002): 440.
41. Thomas Cholmondeley, *Ultima Thule; Or, Thoughts Suggested by a Residence in New Zealand* (London: J. Chapman, 1854), 1; Robert Pemberton, *The Happy Colony* (London: Saunders and Otley, 1854), 25.
42. William Stones, *New Zealand (the Land of Promise) and Its Resources* (London: Algar and Street, 1858), 39.
43. Ernst Bloch, *The Principle of Hope*, vol. 2, trans. Neville Plaice, Steven Plaice, and Paul Knight (Cambridge: MIT Press, 1995), 777.
44. James Belich, *Replenishing the Earth: The Settler Revolution and the Rise of the Anglo-World, 1783–1939* (Oxford: Oxford University Press, 2009), 154. Helen Blythe specifies that this was a decidedly nostalgic utopian literature, in which New Zealand represented "a familiar but new and improved extension of home rather than an entirely foreign destination." Helen Lucy Blythe, *The Victorian Colonial Romance with the Antipodes* (New York: Palgrave Macmillan, 2014), 10.
45. On Bellamy's borrowing from *The Great Romance*, see Dominic Alessio, "The Great Romance, by the Inhabitant," *Science Fiction Studies* 20, no. 3 (1993): 305–40, which also includes the full text of the novella. Lyman Sargent describes New Zealand as "the settler colony with the strongest utopian tradition, with the possible exception of the United States," in "Colonial and Postcolonial Utopias," in *The Cambridge Companion to Utopian Literature*, ed. Gregory Claeys (Cambridge: Cambridge University Press, 2010), 209.
46. Samuel Butler, *A First Year in Canterbury Settlement: With Other Early Essays* (London: A. C. Fifield, 1914), 142.
47. Ibid., 37–44.

48. Ibid., 65.
49. Ibid., 66.
50. Samuel Butler, *Erewhon*, ed. Peter Mudford (London: Penguin, 1985), 257, 76. On Butler's satire of pastoralism, see Philip Armstrong, "Samuel Butler's Sheep," *Journal of Victorian Culture* 17, no. 4 (2012): 442–53.
51. Butler, *Erewhon*, 70. This vista is similar to that seen by Julian West in *Looking Backward*: "At my feet lay a great city. Miles of broad streets, shaded by trees and lined with fine buildings. . . . Public buildings of a colossal size and an architectural grandeur unparalleled in my day raised their stately piles on every side." Edward Bellamy, *Looking Backward, 2000–1887*, ed. Alex MacDonald (Peterborough, Ontario: Broadview Press, 2003), 65.
52. Butler draws this inversion from the Maori institution of "muru," a sanctioned act of plunder directed in some cases against those who had experienced misfortune—as, in an instance recounted by the Irish settler Frederick Maning, when "a man's child fell in the fire and was almost burnt to death." See Frederick Edward Maning, *Old New Zealand* (Auckland: R. J. Creighton, 1863), 106.
53. Butler, *Erewhon*, 135.
54. Samuel Butler, *Life and Habit* (London: Trübner, 1878), 173. Sally Shuttleworth felicitously describes this as "a biological and ontological theory that does not recognize the conventional boundaries of individual identity." "Evolutionary Psychology and *The Way of All Flesh*," in *Samuel Butler, Victorian Against the Grain: A Critical Overview*, ed. James G. Paradis (Toronto: University of Toronto Press, 2007), 148.
55. Butler, *Erewhon*, 202.
56. Ibid., 221.
57. William Empson, *Some Versions of Pastoral* (New York: Norton, 1938), 15.
58. Raymond Williams, *The Country and the City* (London: Chatto and Windus, 1973), 24, 16.
59. Jameson, *Archaeologies of the Future*, xii.
60. "Our Mission," *The Breakthrough Institute*, http://thebreakthrough.org/about/mission (accessed August 17, 2016); John Asafu-Adjaye et al., *An Ecomodernist Manifesto*, http://www.ecomodernism.org/manifesto (accessed August 17, 2016).
61. Edward O. Wilson, *Half-Earth: Our Planet's Fight for Life* (New York: Liveright, 2016), 3, 186.
62. Clive Hamilton, *Earthmasters: The Dawn of the Age of Climate Engineering* (New Haven: Yale University Press, 2013), 104. For a critique of ecomodernism specifically, see Clive Hamilton, "The Theodicy of the 'Good Anthropocene,'" *Environmental Humanities* 7, no. 1 (2015): 233–38.

63. Jason W. Moore, *Capitalism in the Web of Life: Ecology and the Accumulation of Capital* (New York: Verso, 2015), 173.

64. Andreas Malm, *Fossil Capital: The Rise of Steam Power and the Roots of Global Warming* (London: Verso, 2016), 369.

65. Ibid., 383.

66. Christophe Bonneuil and Jean-Baptiste Fressoz, *The Shock of the Anthropocene: The Earth, History and Us* (London: Verso, 2016), 224–25.

CHAPTER 8

From Specimen to System
Botanical Scale and the Environmental Sublime in Joseph Dalton Hooker's Himalayas

Lynn Voskuil

The question of scale has recently become a consequential matter for the humanities, even (or perhaps especially) when it is not explicitly theorized. It motivates the world literature movement, for example, which has reimagined literature as a transnational, even planetary, system that challenges our conceptions of singular texts and defies the conventional moves of close reading.[1] It is likewise implicated in our new, wider sense of the human as linked to other sentient creatures and insentient objects in ontologically variable assemblages that prompt a more ecologically sensitive ethics.[2] In such formulations, scale is often loosely affiliated with "distance" or "extent": We must learn new ways, it is said, to study literature as an infinite, planetary (and not merely a national or even international) phenomenon, and we must broaden our traditional notion of humanity as unique and exceptional—a notion that has inspired the study of literature for centuries. While these are crucial claims, the assumptions that ground much of this scholarship tend to downplay the relations between scales and the process of scaling, the moves that enable both a close focus on and a panoramic view of the objects we study—and the inevitable distortions that ensue. What scales are, why they matter, and how we move among them thus remain underexplored problems for the humanities. As David Palumbo-Liu, Bruce Robbins, and Nirvana Tanoukhi have recently noted, "The gravitational pull of the world scale is clear. What that scale ought to mean to us remains a conundrum."[3]

The consideration of scale is especially important for the branch of humanities that focuses on political and environmental ecologies. While ecological scholarship in the natural sciences recognized the importance of scale long ago, that realization has come more slowly to the humanities and social sciences: Only recently has our study of literature and culture been scaled to a cosmic arena and our awareness of time been expanded from the countable centuries of literary history to the vastness of time implied by the Anthropocene.[4] The challenge is not simply that our objects of study are either small or (newly) large; the challenge is that our very objects and methods of study are themselves transformed by the process of scaling. This essay takes up that challenge by following the lead of ecologists and viewing the concept of scale as a perceptual and hermeneutical problem. Scale may still be a "conundrum" for the humanities because we no longer fully recognize our objects and systems of study, transformed as they now are by new explanatory models. What we see, in other words, "is contingent upon the window through which the system is viewed," as ecologist Simon Levin has put it.[5]

While this intellectual problem is urgently of-the-moment, its antecedents lie in nineteenth-century texts and disciplines that broached the question of scale in variable ways, frequently as a feature of the imperial project. The natural sciences registered the global implications of these issues with particular force because they were central to Britain's imperialist motive and quest. As historian of science Janet Browne has observed, "the study of animal and plant geography in nineteenth-century Britain was one of the most obviously imperial sciences in an age of increasing imperialism."[6] Integral to the imperial project was an ever more complex awareness of the Empire's vast dimensions, an awareness that spurred in turn the development of epistemologies that began to theorize and apply scalar heuristics in a global arena. This essay explores the problem of scale, and its global ramifications, in two volumes by Victorian botanist Joseph Dalton Hooker: *Himalayan Journals* (1854), the narrative of Hooker's midcentury travels through Bengal and the Himalayan mountains collecting plants; and *Flora Indica* (1855), the first installation of a large, systematic botany that aimed to catalogue many of those findings.[7] As one of the most ambitious and renowned plant scientists in the nineteenth century, Hooker was keen to transform the practice of botany from a mere focus on individual specimens to a discipline that also considered global patterns and distribution of plants. With its account of his exploratory travels from 1847 to 1851, *Himalayan Journals* records Hooker's early efforts to grapple with these intellectual problems, while *Flora Indica* systematically explains some of his central intellectual principles. Key to the effect of these volumes was Hooker's mix of rhetorical approaches, approaches we would now characterize as variably "scientific" or "literary" but that he melded to represent plants, his own objects of study. Attention to these now little-read texts brings to the fore many ambiguities of global scale that are still with us today, especially in the humanities.

Hooker was not an early environmental activist. Indeed, as an employee of the British government and a proponent of "economic botany"—the practical study of plants for the purpose of enriching the Empire—he contributed to Victorian experiments that eventually enabled some of the problematic effects we still experience from (for example) massive industrial agriculture. And his intellectual achievements cannot, of course, be separated from either their imperialist motives or their environmental effects: the moral, cultural, and ideological catalysts that propelled the British Empire continue to fuel the kind of "slow violence" that Rob Nixon has discussed so eloquently.[8] At the same time, the effects of Hooker's work are not fully explained by their ideological intentions or contexts. Perplexed by plants that could thrive both in his British garden and in the Himalayan mountains, he began to cultivate an interpretive awareness that could make sense of, and move between, these disparate global regions—a form of awareness that resonates beyond its immediate geopolitical environment. While Hooker did not wholly solve the problems of scale he encountered, he was awake to the hermeneutical uncertainties that emerge when a global consciousness is cultivated. His methods are thus not quite captured by the smooth workings of what Bruno Latour has called the "zoom effect," the cinematic effect that makes movement from the close-up to the panorama seem natural and frictionless.[9] Instead, for Hooker, the shift from an individual object of scrutiny to its global range presented enormous perplexities—and enormous friction. In the mid–nineteenth century, in fact, he anticipated what Timothy Clark has called "scale effects," making this methodological issue an object of study in its own right.[10]

Specimens and Systems

Hooker's involvement with questions of scale has its roots in nineteenth-century conventions of botanic study. In his own era (and earlier), botany—

like the many allied branches of natural history—was both an amateur and professional pursuit, and many of its practitioners were assiduous collectors.[11] As Anne Larsen says, "Natural history in this period was a science based on *specimens*"—on the singular insect, rock, barnacle, taxidermied bird, or, for botanists, the individual plant.[12] Hooker was himself deeply involved in the pursuit and study of individual species and often mired in the detailed minutiae of locating, identifying, recording, and preserving specimens, whether he was collecting them himself or instructing others. As a traveler in his early career, he amassed an impressive herbarium, with plants collected when he was an assistant naturalist on the so-called "Magnetic Crusade" to Antarctica, and then as the lead naturalist on the journey through Sikkim, Nepal, and Bengal that is chronicled in *Himalayan Journals*. Hooker's own writing, moreover, often features a focus on the individual specimen. An important outcome of his Himalayan travels, for example, was his discovery of many species of rhododendron not then known in the West. In 1849, before he returned from his Himalayan trip and with the help of his father back in London (the director of Kew before him), Hooker published *The Rhododendrons of the Sikkim-Himalaya*, a large, lavishly illustrated volume that featured just this kind of species description with the purpose of introducing these new rhododendrons to both botanists and horticulturists (see Figure 2).[13]

Hooker, however, aimed to move beyond mere collection, description, and classification of individual species—although classification in particular remained an important, time-consuming task in an era of rapidly expanding herbaria. Ambitiously, as Jim Endersby has shown, he wanted to elevate botany to a "philosophical" level and transform it into a science that could compete socially and intellectually with chemistry, physics, mathematics, and even astronomy.[14] Only when botanists transcended the local distribution of a species, Hooker argued—only when they could document its potentially global reach—would botany gain a truly systematic foundation and plants be "classified upon philosophical principles" (*Flora* 8).

Hooker was not the first European botanist to scale his study of plants to a global arena. Indeed, Browne's narrative of what she calls "biogeography" begins in the early seventeenth century, with theologians who sought to explain, often by interpreting Genesis literally, how animals exiting Noah's ark repopulated the entire earth. The recognition of exotic species, even in the seventeenth century, prompted an awareness of geographical regions around the world that could accommodate the different survival needs of variable flora and fauna, an issue that natural historians also took

Figure 2. *Rhododendron Dalhousiae*, an example of a species description provided by a single specimen (1849–51). (Image from Joseph Dalton Hooker, *The Rhododendrons of the Sikkim Himalaya*, reproduced by kind permission of the Board of Trustees of the Royal Botanic Gardens, Kew.)

up. By the end of the eighteenth century, however, as reliance on the Noachian story declined, plants assumed a more significant position than animals in emerging studies of global migration and distribution. Literally rooted in the soil, plants were tied more closely to the physical environment, thereby serving as more reliable biogeographical markers.[15] Late-eighteenth-and early-nineteenth-century plant scientists thus played an important role in the development of global study. Joseph Banks, Hooker's eighteenth-century predecessor at Kew, was himself an intrepid explorer who bolstered Kew's collections of exotic specimens and worked with King George III to establish the botanical garden as a center of agricultural "improvement" in the service of empire.[16] Banks, however, produced few significant writings and was known more as an entrepreneur and manager than as an important botanist; while he expanded Britain's scientific networks, he spent little time pondering global theory.[17]

Although Hooker learned administrative lessons from Banks's tenure at Kew, he was inspired intellectually (if sometimes indirectly) by scientists who studied global plant migration and distribution in the early nineteenth century. Perhaps chief among these was Alexander von Humboldt, the German explorer-philosopher-scientist and prolific author whose wide influence in nineteenth-century Europe, Britain, and America is difficult to over-state; Susan Cannon, in fact, has described the first half of the nineteenth century overall as the era of "Humboldtian science."[18] Among Humboldt's many innovations was the invention of the isobar, a cartographic line that links points with the same atmospheric pressure around the world, and the isotherm, a similar device that joins points with the same minimum or maximum average temperatures; both became mechanisms for the transaction of scientific study on a global scale. Also influential were the French scientist Augustin-Pyramus de Candolle and (especially for Hooker) the British botanist Robert Brown.[19] Hooker's advances paralleled and built on the work of these earlier botanists, but he also crucially circulated the central premise of distribution studies: the conviction that botany should be studied from a global perspective. Himself a prolific correspondent and collaborator, Hooker discussed his ideas with many other scientists—including his close friend Charles Darwin (to whom the *Himalayan Journals* are dedicated)—thereby extending the reach of global thought in British intellectual circles and, through his position at Kew, in the wider culture as well. Revered then as one of the most important nineteenth-century botanists, Hooker developed and spread methodologies that linked global botanic communities as well as the scalar heuristics that were implied by such methods.

With their emphasis on both global plant communities and species classification, Hooker's goals for botany entailed principles of study that necessitated processes of scaling, processes that move from observation of the individual specimen to analysis of its distribution around the world and that involve the inference of relations among different spatial scales. As Benjamin Morgan has observed, scales "structure perception... by defining levels of granularity."[20] For centuries, botanists had carried out their study largely by viewing a single specimen closely, on a very localized scale: by counting and labeling its parts and viewing them microscopically, by comparing it to other specimens, even by touching its leaves and stems. While Hooker used these observational practices, he didn't fully trust the authority even of his own, well-trained eyes—a posture of skepticism that testifies to his excellence as a botanist. As Lorraine Daston has shown, nineteenth-century botanists looked to "type specimens" for classificatory purposes. Whether they existed as desiccated herbarium specimens or botanical illustrations (see Figure 2), these types were supposed to exemplify the species neither naturalistically (as in the field) nor idealistically (as the one essential specimen); instead, type specimens were seen as the distillation of many examples of the species.[21] The individual example of a species—in the field or elsewhere—was thus not interpreted in its own singularity; indeed, its unique qualities, often notable to gardeners and horticulturists, were seen as radically insufficient, even detrimental, to the task of accurate classification. The solution to such finely scaled botanic myopia, then, was itself a solution of scale: While close sensory inspection was an essential element of accurate classification, the observed plant must be placed within a much larger class of similar examples and one scaling mechanism be used instead of—or in addition to—another.

Hooker argued, however, that plants should be studied not only as individual species but in larger, even global aggregates. In *Flora Indica*, he states the principle at stake: "It will be generally found that botanists who confine their attention to the vegetation of a circumscribed area, take a much more contracted view of the limits of species, than those who extend their investigations over the whole surface of the globe" (13). But an individual botanist could obviously not personally inspect all existing specimens—or even representative specimens—on a world-wide basis. How, then, could the study of plants be made globally systematic? Hooker addressed this question by taking a methodological cue from Humboldt: He emphasized the frequent recording of measurements and data. Such measurements are a significant representational feature even of *Himalayan Journals*, a volume that targeted a wide audience (in distinction from *Flora Indica*, which assumed a more specialized readership). While the *Journals* echo other nineteenth-century travel genres by offering descriptions of the people and customs he encountered along the way, the inclusion of calculations and figures is both insistent and persistent. A typical example occurs deep into the first volume: "The mean temperature of the twenty-four hours was 32 7 (max. 41 5/ min. 27 2), mean dew-point 29.7, and saturation 0.82.... The black bulb thermometer rose to 132, at 9 a.m. on the 28th, or 94 2 above the temperature of the air in the shade" (*Journals* 1.310–11). This quotation, excerpted from a much longer paragraph that provides many additional figures, exemplifies a commonplace of this text. Far from an incidental stylistic tic, these calculations are important because they register Hooker's effort to make use of different scales. Relationships among scales were significant for him, and the question of how to move among them was a vexing problem.

Hooker's use of data helped him address this problem. In data, Hooker and other botanists such as Humboldt and Brown—as well as the many other nineteenth-century scientists who shared their convictions about the power of data—found a tool that enabled them to chart global phenomena, even those they couldn't observe firsthand. Jen Hill has recently illuminated the significance of such pursuits with her analysis of "correlation" in Francis Galton's study of the "thermodynamics of weather." As Hill argues, Galton's work revealed "correlation to him—the ways in which local conditions were part of larger systems, determined and thus explained by literally invisible forces and geographically distant patterns."[22] Hooker, it might be said, anticipated Galton's attraction to "correlation" by using the data he recorded in his travels to represent what he calls, in *Himalayan Journals*, "zones" and "belts" of plant distribution (*Journals* 1:142–43 and 1:348–49). He explores such concepts more extensively in *Flora Indica* by theorizing a region he called the "botanical province" and dividing the ground covered by his *Flora* into dozens of such provinces (88). Hooker found this division process to be "a very much more difficult task than might have been supposed" because the lines he drew conformed to "physical features rather than arbitrary lines" or "political boundaries" 84, 88). His botanical provinces were circumscribed, in other words, by the botanic, climatic, geological, and geographical measurements he had so carefully recorded throughout his travels.

These provinces, and the challenges of delineating them, play a significant ideological role in Hooker's work. In charting them, first of all, he ignored the current boundaries of British possessions in the South Asian continent, surveying instead a region that was much larger. As an employee of the British government, Hooker was paid to explore territory, plants, and scientific data that would benefit the Empire—especially, in this case, its Indian colonies. In and of itself, the act of redrawing regional boundaries could be construed as problematic from the standpoint of empire. Although Hooker's intention was certainly not to subvert government objectives, his indifference to the boundaries of British possessions, and his use of physical data to chart his provinces, implicitly challenged the criteria that dictated the establishment of territorial borders.[23] Beyond this geopolitical provocation, Hooker's provinces also tested the limits of current ideologies that governed the study of botany, a provocation he took up more explicitly. Aware of potential controversy, he devoted several pages in *Flora Indica* to defending the breadth of its geographical reach and rationalizing the provincial borders he outlined (83–90). One of his stated goals—both for *Flora Indica* and for the study of plants more generally— was to "banish prejudice from the domain of Systematic Botany" by demonstrating the enormous geographical range of many botanic species (*Flora* 88). This range not only ignored political boundaries but also directly countered the common practice, widespread among practical nineteenth-century botanists, to proclaim a new species when an unfamiliar plant was discovered and thereby, according to Hooker, falsely inflate the total worldwide number of species, a practice he deplored as "*hair-splitting*" (*Flora* 13). He believed that even the study of plants was informed by "a proneness of the human mind to regard everything from an unknown country, or that is seen surrounded with foreign association, as itself unknown" (*Flora* 87). Instead of perceiving a new species each time he encountered an apparently unfamiliar plant in a new region, Hooker assumed that it probably existed elsewhere in the world; and he then searched for it, using his own observations and collections, the work of other botanists, and the physical data he recorded and charted. The resulting botanical zones traversed national and even some natural boundaries (mountain ranges, for example), drawing the plant world together in "large cosmopolitan families" that rambled far beyond the localized species groups identified by many botanists (*Flora* 90).

Visual representations of these zones differ strikingly from more conventional representations of plants. The lithographs in Hooker's opulent rhododendron volume exemplify what had become traditional botanical illustration by the mid–nineteenth century (see Figure 2); the influence of Linnaean taxonomy is reflected in a close-up of the flower, with reproductive parts drawn separately, enabling a granular, finely scaled focus on a single specimen.[24] In *Flora Indica*, although Hooker devoted many pages to verbal species descriptions, he included no such specimen drawings, offering instead only two cartographic images. While one map shows the physical geography of India, the other represents isothermal belts that loop across several nations and continents, including South Asia, East Asia, Africa, and Europe (see Figure 3). Originating, as Hooker noted, in *On the Distribution of Heat over the Surface of the Globe* (1853), a recently published work by pioneering climatologist Heinrich Wilhelm Dove translated by Elizabeth Sabine, this map displays the kind of physical data that corresponded to the plant distributions Hooker charted (*Flora* 258). It thereby graphs the potential range of a so-called "English" plant that might also be found in, say, India, linking both plants and data into large-scale aggregates that ignored national borders.

The zones represented by the isothermal map and the measurements typified by the figures Hooker inserted prolifically into *Himalayan Journals* could presumably have conferred the certainty and respect he sought for the pursuit of a globalized, "philosophical" botany. As Mary Poovey has argued, the practices of counting, measuring, and figuring had assumed a new primacy in the late eighteenth and early nineteenth centuries, a shift that gave such data a new credence and authority derived from their ostensible neutrality. Hooker's use of botanic zones and belts could even be seen as a partial response to what Poovey calls "the problem of induction" because, as she puts it, "measuring, counting, and figuring proportions and variations would bridge the gap between the observed particular and general knowledge."[25] Maps and tables of physical data that arranged global plant communities in "cosmopolitan" families might be used in this way, deployed as a mechanism to move from the single plant specimen to their world-wide distribution. Hooker did not make this argument explicitly in either *Himalayan Journals* or *Flora Indica*, however. Instead, he acknowledged the frequently difficult task of making scalar moves from a single specimen or individual site to global plant communities. His *Himalayan Journals* are especially important in this regard. While the more systematic *Flora* sets forth his intellectual principles and ambitions for botanic study, the less conclusive *Journals* includes his on-the-spot ruminations and reflections on what he was discovering and collecting. Indeed, the very form of the journal as a genre—a form whose generic conventions are often used to note and work through intellectual or emotional ambiguity—enabled Hooker to voice equivocations and skepticism about his initial findings. The *Journals* are thus a useful record of his perplexities about how scale works in the field. Notably, he conveys these perplexities not by

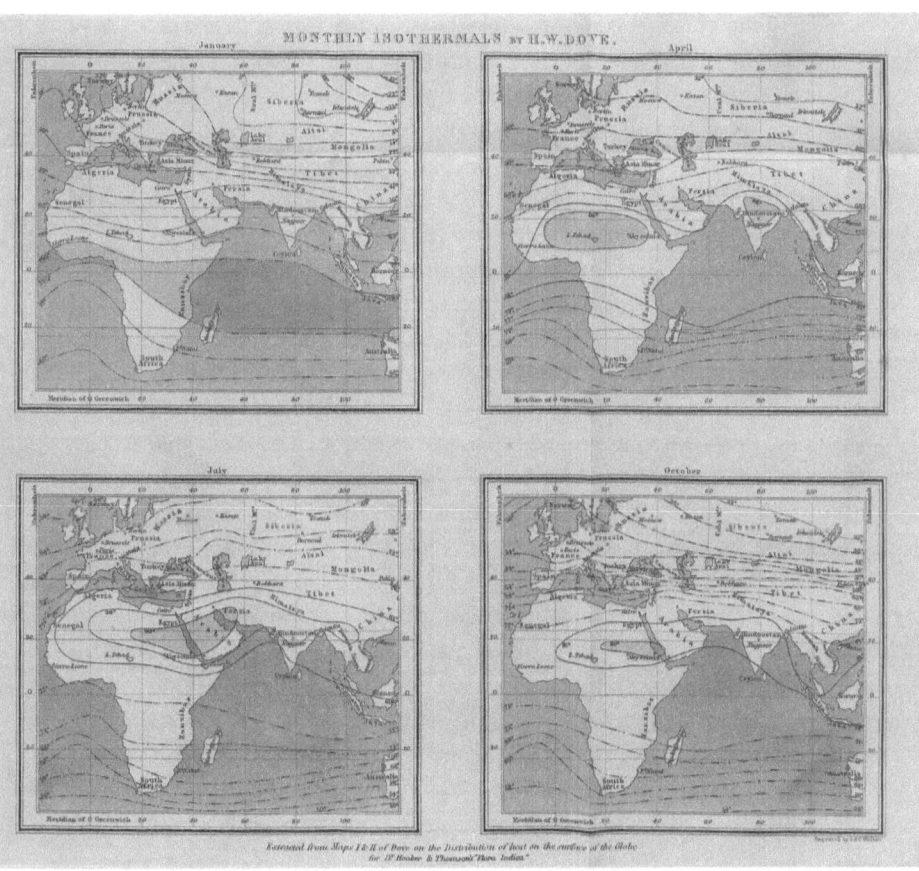

Figure 3. Isothermal zones, one of two illustrations from *Flora Indica* (1855). (Image reproduced courtesy of Biodiversity Heritage Library.)

reworking or questioning his data but by adapting the aesthetic conventions of the sublime and the perspectival framework of landscape.

Scalar Distortions

These conventions, and their relations to scale, are introduced early in *Himalayan Journals*, when Hooker first notes a radical zonal shift in physical phenomena. He is in the Himalayan foothills, having proceeded northward from Calcutta (Kolkata) through the states of West Bengal and Bihar. "Every feature, botanical, geological, and zoological, is new on entering this district," he notes. No "botanical region [is] more clearly marked than this, which is the commencement of Himalayan vegetation" (*Journals* 1.100). Commenting during his first ascent into the Himalayas proper, he expresses his amazement at the size of the plant and animal species he was encountering. "Upon what a gigantic scale does nature here operate!" he marvels (*Journals* 1.104). "The prevalent timber is gigantic," he continues, "and scaled by climbing *Leguminosae*, as *Bauhinias* and *Robinias*, which sometimes sheath the trunks, or span the forest with huge cables, joining tree to tree. Their trunks are also clothed with parasitical Orchids, and still more beautifully with Pothos (*Scindapsus*), Peppers, *Gnetum*, Vines, Convolvulus, and *Bignoniae*" (*Journals* 1:108). Hooker here evokes some apparently standard meanings of scale. In the tropical landscape, first of all, plants scale each other: Vines use larger, more stable species as ladders or steps to reach the sunlight. Both flora and fauna, moreover, vastly exceed their counterparts in the temperate climates Hooker knew best; not only do plants grow taller and reach further, but they are far more dense and profuse than the plants in European landscapes.

While fairly straightforward, these intimations of scale are more complex and allusive than their uses here might initially suggest. Hooker's descriptions of tropical forests near the Himalayas once again echo those of Humboldt, most notably those included in his *Personal Narrative of Travels to the Equinoctial Regions of the New Continent, 1799–1804*. Humboldt's evocations of tropical landscapes in that work had become something of a convention in naturalist writing and themselves drew on turn-of-the-century evocations of the sublime and the picturesque.[26] As Humboldt wrote, for example, "everything is gigantic" in the forest of South America, "the mountains, the rivers, and the mass of vegetation.... The trunks of the trees are every where concealed under a thick carpet of verdure...."[27] Hooker draws on similar aesthetic conventions. "Dissolving views gives some ideas of the magic creation and dispersion of ... effects," he notes, "but any combination of science and art can no more recall the scene, than it can the feelings of awe that crept over me, during the hour I spent in solitude amongst these stupendous mountains" (*Journals* 1.266). The evocation of sublimity and the echo of Humboldt in *Himalayan Journals* not only place Hooker in a nineteenth-century tradition of botanical writing but also show him reaching for available aesthetic means to grasp the scale of what he was observing in the Himalayas.[28]

It is no surprise that Hooker used familiar aesthetic conventions to describe Himalayan scenes. In preceding decades, of course, Romantic poets had often evoked the sublime to capture their own responses to nature, with mountain prospects figuring significantly as triggers of awe or terror. A long, subsequent critical focus on the literary and philosophical sublime—too lengthy and complex to unpack here—reached something of a consensus in the late twentieth century that the sublime had been individuated, internalized, and (often) domesticated in the nineteenth century, eventuating in a concept of sublime nature that came to be "nothing without the mind's imaginings," as David Simpson has put it.[29] This philosophical correlation of sublimity with human subjectivity has itself been affiliated with the imperialist project, as Simpson also notes. The "coincidence" of an egotistical sublime "with the expansion of empire and capital," he writes, "might cause us to suspect that there is something ethically uncomfortable at the heart of our craving for bigness and our urge to set ourselves against enormity in a process of cognizance or conquest, whether of depth, space, or territory."[30] In a related trope of territorial mastery—the vision of the imperial explorer metaphorically conquering the sublime scene he surveys by aestheticizing it—Mary Louise Pratt identified a potent, Victorian version of this urge and found it embodied in the figure of the explorer Richard Burton.[31]

Hooker's evocations of sublimity in *Himalayan Journals* thus come with a heavy load of ideological freight. His uses of the sublime, however, while recognizably rooted in their era, attenuate some of these ideological associations. Consider the quality that Edmund Burke had called "vastness" or "greatness of dimension," a "powerful cause of the sublime" featured in his *Enquiry into the Sublime and Beautiful*.[32] Hooker predictably emphasizes this quality in his descriptions of the Himalayas, echoing other writers who had made similar observations about the European Alps. Provocatively, though, he often disrupts this emphasis by reframing even the most marvelous Himalayan spectacles with remarks about scalar distortions that cloud the viewer's perceptions. He tackles these circumstances immediately in the preface, where he discusses the landscape illustrations included in the volume, which were made from his own sketches (see Figure 4). Observing that his drawings would be considered "tame" if compared with those of landscape artists who exaggerate certain features, he adds that "the total effect of steepness and elevation, especially in a mountain view" could not be accurately represented in a "small scale," accurately proportioned sketch (*Journals* xviii).

This theme of scalar distortion is sustained throughout the text no less insistently than the emphasis on measurement and often in scenes that are framed as sublime. Chapter 5, for example, opens with the disclaimer, conventional in representations of sublimity, that these "sublime phenomena"—his first view of the highest, snow-clad elevation in the Himalayas—"elude

Figure 4. A lithograph of the Himalayan landscape, probably based on a sketch made in the field by Hooker (1854). (Image from Joseph Dalton Hooker, *Himalayan Journals*, reproduced courtesy of Biodiversity Heritage Library.)

all attempts at description." Also conventionally, Hooker then proceeds to describe them nonetheless, including the "sensations and impressions that rivet" him to the scene (*Journals* 1:123). But he follows these conventions with a series of figures that record latitude, elevation, and the angles of peaks with respect to the horizon. The latter, taken with a theodolite (a surveying instrument), are far smaller than expected and clash sharply with the apparent enormity perceived by the naked eye (*Journals* 1:124). For Hooker, sheer size was only part of it; proportions and relations of scale are equally at stake here, no less than immensity.

In this case, the use of equipment seems to address the ambiguities of scalar distortion, thus elevating the scientist to a position of cognitive conquest: like Pratt's Burton, Hooker uses Western knowledge to survey the scene and shrink it down to size. In several other cases, however, doubting even his own instruments, Hooker implies that the scale of Himalayan nature is almost beyond the measure of Western perception. One such episode is especially illuminating, noted when he was looking northward toward Tibet and westward toward Nepal from an elevation of 20,000 feet in the Sikkim Himalayas. "This wonderful view," he remarks, "forcibly impressed me with the fact, that all eye-estimates in mountainous countries are utterly fallacious, if not corrected by study and experience." After a lengthy description, he concludes with a hermeneutical image. "The want of refraction to lift the horizon, the astonishing precision of the outlines, and the brilliancy of the images of mountains reduced by distance to mere specks," he marvels, "are all circumstances tending to depress them to appearance. The absence of trees, houses, and familiar objects to assist the eye in the appreciation of distance, throws back the whole landscape; which, seen through the rarified atmosphere of 18,500 feet, looks as if diminished by being surveyed through the wrong end of a telescope" (*Journals* 2:127–28). As Hooker attempts to make sense of the panorama in front of him, he reaches instinctively for tools of his trade that he can use to measure it—to gauge, calibrate, and enumerate its features and to generalize from these data, as he had when he designed his botanical provinces. In this case, however, there are no humanized elements of landscape—no "trees, houses, and familiar objects"—that can be used to scale his view in this way, with the result that his instrumental grasp of the surroundings is shaky at best. Notably, even the trusty telescope has become unreliable, a tool that now distorts the scene rather than clarifying it. Here, in other words, even "study and experience" are not definitive. In fact, by shrinking the scene, the telescope paradoxically magnifies the contortions of the scale effect, an effect so potent that it can shrink a global entity—the highest mountain in India, say—into a pocket-size landscape. At this point, far into his journey, Hooker frames himself and the lofty scene with obscurity rather than clarity, as if to underscore its strangeness once more, just as he did when he first began to ascend the Himalayan foothills.

Throughout *Himalayan Journals*, Hooker frequently reaches for the language of the sublime when he is faced with the vastness of his methodological challenge, which includes the shifting scales on which his work must be carried out if botany is to become "philosophical." At such moments, the authority of his own senses, measurements, or experience appears diminished. Prevailing in this and many related scenes in the *Himalayan Journals* is an emphasis on the immeasurable mysteries of the surrounding atmosphere and biosphere, the sense that their properties exceed the capacities of even the scientific mind to comprehend them. Rather than verifying his scientific subjectivity, then, the Himalayan panoramas often undermine it, exemplifying what Emily Brady has called the "the environmental sublime," an ontological state characterized by "an overwhelming of the subject, in which the self is dislocated through a sense of nature as mysterious, and neither fully known nor appropriated by human reason."[33] In distinction from the egotistical sublime that Simpson (and Keats) discusses, Brady's notion destabilizes both the subject's self-awareness and the knowability of the natural scene, including the subject's own position within it. For Hooker, the Himalayan mountains were not easily conquered in aesthetic, cognitive, or imperial terms. Instead, their very presence—their vastness, their inscrutability, their indifference to geopolitical boundaries—registered the limits of his politics, his senses, and even his science.

That Hooker frames the sites of his bafflement as themselves framed is significant for several reasons.[34] His decision, first of all, to frame them as sublime underscores the difficulty not merely of grasping the scenes but of representing them. When he evokes the sublime, he takes advantage of an aesthetic vernacular familiar to his British audience, a language that humanizes his scientific findings in ways that his readers will fathom. This very familiarity, however, guarantees that his uses of the sublime—a set of tropes that was well-worn by mid-century—will call attention to themselves as representational tools. And they do so precisely at that point when the environmental sublime makes itself felt, when the scientist is confronted with the unfathomability of his natural object of study. This incongruence—the use of highly conventionalized language when faced with a profoundly unconventional vista—underscores the practice of globalized study as a hermeneutical enterprise. By using tropes of the sublime, Hooker suggests that even the most natural, untouched sites and objects imaginable must be mediated by cultural and methodological mechanisms. Secondly, however, his work also demonstrates the limits and constraints of the very mechanisms it deploys. If readers miss the significance of the sublime frameworks in *Himalayan Journals*, they might more readily grasp the framing functions of his instruments and measurements: the theodolite that makes the highest mountains in the world seem small, the telescope that shrinks entire panoramas. On the one hand, these instruments challenge sensory evidence — ocular testimony, for example, unsupported by scientific aids. On the other, they remind both Hooker and his readers that scientific instruments, like the human eye, are themselves lenses that leave certain elements out of the picture. With a theodolite, you can measure the angles of distant mountains with respect to the horizon, but there is not much it can say about the botanic specimen at your feet.

In the end, though Hooker made many important advances in the global study of plants, he did not fully solve the methodological and, especially, the hermeneutical issues he raised. This condition of his work perhaps explains why there is a lingering sense of "the environmental sublime" in both of his mid-century texts but most notably in *Himalayan Journals*— the sense that natural objects of study, whether small or enormously large, cannot be fully grasped even with sophisticated instruments. This conundrum speaks to conceptions of scale both now and in the nineteenth century. Today, perhaps building on Hooker's now distant legacy, ecologists explicitly recognize that "there is no natural level of description," as Levin puts it; scales are observational tools that necessarily focus on variable kinds of information.[35] This recognition, seemingly leaves little room for an environmental sublime as Brady explains it: All natural phenomena are scientifically explicable, if not currently then eventually. Instead of relinquishing apparently natural phenomena to a quasi-metaphysical realm, "the key," as Levin says, is to recognize scalar variability, "to determine what information is preserved and what information is lost as one moves from one scale to the other."[36] One hundred and fifty years earlier, Hooker was awake to the ambiguities of these interpretive and scalar questions and brought them to the attention of fellow botanists and popular readers even as he furthered his scientific agenda. Rather than using aesthetic tools to invalidate quantitative results, however, or drawing on exclusively quantitative tools to solve his interpretive problems, Hooker employed both representational modes to highlight the processes of scaling and to carry out the difficult epistemological task of moving from the observation of the local, individual object of study to the development of global, systematic knowledge.

The provenance of this critique in nineteenth-century science remains provocative for the humanities today. Our turn to the question of quantity—"that forgotten 99 percent" of nineteenth-century British novels, as Franco Moretti puts it[37]—is an important, even crucial, turn. At the same time, as Hooker's work exemplifies, quantitative methods are not, in and of themselves, sufficient, nor is neglect of the individual object of study a prerequisite for global analysis of various kinds, including distant reading. If this sounds like a retrograde move, a return to the imperatives of close reading, it is not meant to be. Instead, drawing on Hooker's analogy, it is meant to emphasize the epistemological challenges that are endemic to new modes of studying the humanities on a global scale or as a world-wide system. As Hooker demonstrated in his own discipline, these challenges are not simple questions of breadth, size, or countability; they are instead hermeneutical questions that require multiple disciplinary frames and methods to answer, including the interpretive and aesthetic methods in which humanists are trained. While books and plants are not fully interchangeable, ecosystems have certain affinities with world systems of literature. How we perceive those systems and construct their variable scales of study is a question of method that is as central to the humanities as it is to the natural sciences, both now and in the nineteenth century.

Notes

1. See, for example, Pascale Casanova, *The World Republic of Letters*, trans. M. B. Debevoise (Cambridge: Harvard University Press, 2004); and Franco Moretti, *Distant Reading* (London: Verso, 2013).

2. For important contributions to this discussion, see Jane Bennett, *Vibrant Matter: A Political Ecology of Things* (Durham: Duke University Press, 2010); Dipesh Chakrabarty, "Postcolonial Studies and the Challenge of Climate Change," *New Literary History* 43, no. 1 (2012): 1–18.

3. David Palumbo-Liu, Bruce Robbins, and Nirvana Tanoukhi, Introduction to *Immanuel Wallerstein and the Problem of the World: System, Scale, Culture*, ed. David Palumbo-Liu, Bruce Robbins, and Nirvana Tanoukhi (Durham: Duke University Press, 2011), 4. While scale remains an unresolved issue for the humanities, certain scholars have begun to address the problem. Timothy Morton, for example, has explored some of its spatial and temporal dimensions. See in particular *Hyperobjects: Philosophy and Ecology after the End of the World* (Minneapolis: University of Minnesota Press, 2013).

4. Among the many, *many* discussions of scale in the discipline of ecology, see Simon A. Levin, "The Problem of Pattern and Scale in Ecology,"

Ecology 73, no. 6 (1992): 1943–67; and Dean L. Urban, "Modeling Ecological Processes across Scales," *Ecology* 86, no. 8 (2005): 1996–2006.

5. Levin, "Scale," 1953.

6. Janet Browne, "Biogeography and Empire," in *Cultures of Natural History*, ed. N. Jardine, J. A. Secord, and E. C. Spary (Cambridge: Cambridge University Press, 1996), 305.

7. Joseph Dalton Hooker, *Himalayan Journals; Or, Notes of a Naturalist in Bengal, the Sikkim and Nepal Himalayas, the Khasia Mountains, &c*, 2 vols. (London: John Murray, 1854); Joseph Dalton Hooker and Thomas Thomson, *Flora Indica: Being a Systematic Account of the Plants of British India* (London: W. Pamplin, 1855). Hereafter both texts will be cited parenthetically within the text, with shortened titles.

8. Rob Nixon, *Slow Violence and the Environmentalism of the Poor* (Cambridge: Harvard University Press, 2011).

9. Bruno Latour, *Reassembling the Social: An Introduction to Actor-Network-Theory* (Oxford: Oxford University Press, 2005), 186; see also 183–90.

10. Clark describes "scale effects" as the "jumps and discontinuities" between scales that are obscured by smooth in-and-out zooming. "Scale," *Telemorphosis: Theory in the Era of Climate Change*, vol. 1 (Ann Arbor: Open Humanities Press, 2012), par. 4. Derek Woods offers the phrase "scale critique," which also captures aspects of Hooker's thinking, in "Scale Critique for the Anthropocene," *Minnesota Review* 83 (2014): 133.

11. My use of the word "professional" here is more or less in keeping with our own understandings today. As Jim Endersby has shown, however, the word "professional" was not necessarily how Hooker preferred to have his own work, or person, described. *Imperial Nature: Joseph Hooker and the Practices of Victorian Science* (Chicago: University of Chicago Press, 2008), 21–22.

12. Anne Larsen, "Equipment for the Field," in *The Cultures of Natural History*, ed. N. Jardine, J. A. Secord, and E. C. Spary (Cambridge: Cambridge University Press, 1996), 358.

13. Joseph Dalton Hooker, *The Rhododendrons of Sikkim-Himalaya*, ed. Sir W. J. Hooker (London: Reeve, Benham, and Reeve, 1849).

14. On Hooker's collecting and preparation of specimens, see Endersby, *Imperial Nature*, Chapter 2; on Hooker's ideas about botany as a "philosophical" and "systematic" science, see Ibid., Chapter 9.

15. Janet Browne, *The Secular Ark: Studies in the History of Biogeography* (New Haven: Yale University Press, 1983), 1–57.

16. For an account of Banks's imperial activities at Kew, see Richard Drayton, *Nature's Government: Science, Imperial Britain, and the 'Improvement' of the World* (New Haven: Yale University Press, 2000), 85–128.

17. On this point, see David Philip Miller, "Joseph Banks, Empire, and 'Centers of Calculation' in Late Hanoverian London," in *Visions of Empire:*

Voyages, Botany, and Representations of Nature, ed. David Philip Miller and Peter Hans Reill (Cambridge: Cambridge University Press, 1996), 21–22.

18. Susan Faye Cannon, *Science in Culture: The Early Victorian Period* (New York: Dawson and Science History Publications, 1978), 73–110. See also Michael Dettelbach, "Humboldtian Science," in *The Cultures of Natural History*, ed. N. Jardine, J. A. Secord, and E. C. Spary (Cambridge: Cambridge University Press, 1996), 287–304.

19. The detailed ideas of Humboldt, de Candolle, Brown, and related botanists on these issues are discussed in Browne, *Secular Ark*, 42–57, and Endersby, *Imperial Nature*, 235–243.

20. Benjamin Morgan, "*Fin du Globe:* On Decadent Planets," *Victorian Studies* 58, no. 4 (2016): 13.

21. Lorraine Daston, "Type Specimens and Scientific Memory," *Critical Inquiry* 31 (2004): 153–82.

22. Jen Hill, "Whorled: Cyclones, Systems, and the Geographical Imagination," *Nineteenth-Century Contexts* 36, no. 5 (2014): 447.

23. At one point, in fact, that indifference led him and his traveling companion into disputed territory in Tibet and a brief imprisonment. See Hooker, *Journals*, 2:190–24.

24. See Gill Saunders, *Picturing Plants: An Analytical History of Botanical Illustration* (Berkeley: University of California Press, 1995), 88–100.

25. Mary Poovey, *A History of the Modern Fact: Problems of Knowledge in the Sciences of Wealth and Society* (Chicago: University of Chicago Press, 1998), 286.

26. On naturalists' use of familiar Romantic conventions of the sublime and picturesque, see Luciana L. Martins, "A Naturalist's Vision of the Tropics: Charles Darwin and the Brazilian Landscape," *Singapore Journal of Tropical Geography* 21, no. 1 (2000): 20; and Lynn Voskuil, "Sotherton and the Geography of Empire: The Landscapes of *Mansfield Park*," *Studies in Romanticism* 53, no. 4 (Winter 2014): 604–8.

27. Alexander von Humboldt and Aimé Bonpland, *Personal Narrative of Travels to the Equinoctial Regions of the New Continent, During the Years 1799–1804*, 2nd ed., trans. Helena Maria Williams (London: Longman, Hurst, Rees, Orme, and Brown, 1822), 3:36.

28. David Arnold discusses the influence of both Humboldt and Darwin on Hooker's style in "Envisioning the Tropics: Joseph Hooker in India and the Himalayas, 1848–1850," in *Tropical Visions in an Age of Empire*, ed. Felix Driver and Luciana Martins (Chicago: University of Chicago Press, 2010), 150–51.

29. David Simpson, "Commentary: Updating the Sublime," *Studies in Romanticism* 26, no. 2 (1987): 255.

30. Ibid., 246.

31. Mary Louise Pratt, *Imperial Eyes: Travel Writing and Transculturation* (London: Routledge, 1992), 201–8.

32. Edmund Burke, *A Philosophical Enquiry into the Origin of our Ideas of the Sublime and Beautiful*, ed. James T. Boulton (Notre Dame: University of Notre Dame Press, 1958), 72.

33. Emily Brady, "The Environmental Sublime," in *The Sublime: From Antiquity to the Present*, ed. Timothy M. Costelloe (Cambridge: Cambridge University Press, 2012), 180.

34. Morgan also discusses scales and frames, with great subtlety and nuance, and an emphasis different from mine here, in *"Fin du Globe,"* 613–14.

35. Levin, "Scale," 1947.

36. Ibid., 1950.

37. Moretti, *Distant Reading*, 67.

CHAPTER 9

"Infinitesimal Lives"
Thomas Hardy's Scale Effects
Aaron Rosenberg

"London. Four million forlorn hopes!" reads Hardy's notebook entry for April 5, 1889.[1] Two days later, still contemplating the despair amassed in what was then the world's most populous city, Hardy wrote, "A woeful fact—that the human race is too extremely developed for its corporeal conditions, the nerves being evolved to an activity abnormal in such an environment." The strain that London's overdeveloped urban environment clearly exerted on Hardy's own "nerves" opens onto a wider lament about the misery of the "human race," which must suffer collectively the biological burden of consciousness. "Even the higher animals are in excess in this respect," he continues, dilating the already immense range of these speculations by questioning "whether Nature, or what we call Nature, so far back when she crossed the line from invertebrates to vertebrates, did not exceed her mission."[2] "This planet," Hardy concludes, "does not supply the materials for happiness to higher existences. Other planets may, though one can hardly see how."

Hardy's use of the word "excess" refers both to his scientific understanding of the world's processes and to an embodied sense of disproportion produced by the enormous measurements of time and space included in that understanding. A kind of rhetorical excess reflects the magnitude of his outlook as well, and we may be tempted to suggest that Hardy goes too far by turning the sympathetic faculties of his own "too extremely developed" nervous system to phenomena that seem oblivious to, and incommensurable with, daily lived experience.[3] From the aggregated mood of London's burgeoning population, to the moment in deep time when "Nature" first transgressed, to the material possibilities for life on other planets—the scale of these concerns seems to promote an excessive mode that may strike us as "unrealistic."

As critics have observed, Hardy's efforts to orient the lives of individuals within what Pamela Gossin calls his "personal construction of an astronomical-literary cosmology" dramatize an epistemological crisis: how to regard the categorical significance of "the human."[4] Gillian Beer articulates this as "the problem of finding a scale for the human," a challenge for late Victorians who, in the wake of evolutionary theory, began looking for "a scale that will be neither unrealistically grandiose, nor debilitatingly reductive, which will accept evanescence and the autonomy of systems not serving the human, but which will still call upon Darwin's often-repeated assertion: 'the relation of organism to organism is the most important of all relations.'"[5] We might build on Beer's formulation by recognizing that the literary implications of "systems not serving the human" are not just philosophical or ethical, but deeply formal. If the mimetic effects of the novel—especially the realist novel—depend on detailed representations of human perspectives, empirically observed, then including the scales of "systems not serving the human" introduces a kind of narrative excess, a surplus that threatens to disrupt the novel as a system of human relations, to derange the conventions that give it form, and make it, too, seem "unrealistically grandiose or debilitatingly reductive."

For Hardy, human experience could not be extricated from its situation within scales that spanned geologic time and cosmic distances; however, accommodating this massively distributed reality within a form grounded in everyday life attenuated narrative conventions such as character, setting, and plot—producing "scale effects."[6] As Timothy Clark has demonstrated, "scale effect" is a term of art in structural engineering that can be applied more broadly to forms, including literary ones, whose integrity depends on proportional consistency. Making a form larger or smaller simply by scaling its dimensions up or down is not a straightforward process because a drastic change of degree can cause a change in *kind*. Shifts of scale can expose forms to unsuspected forces that obtain only at certain sizes, or which vary greatly between them, causing destabilizing or disruptive outcomes.[7]

This essay will argue that radical changes of scale tend to occasion shifts of narrative mode: Specifically, fluctuations between a realist mode of narration and a heightened mode of "excess" associated with nonrealist genres, such as sensation fiction, melodrama, and romance.

The notebook entry introduced at the beginning of this essay offers a brief illustration: Tracing the collective unhappiness of "higher existences" to the biological genesis of consciousness—an inscrutable, unintentional event from the deep past—Hardy shifts into a romantic mode by invoking a feminized figure of "Nature" (another form of higher existence) in the act of erring from her intended plot, exceeding her "mission." In other words, to grasp the magnitude of this scientific reality in terms of human experience, Hardy ascribes an almost mythological agency and narrative purpose to nonhuman systems. While this juxtaposition of the scientific and the tropological implies a "re-enchantment" of the rationalized, secular world, it is also indicative of a technique of formal accommodation, a means of channeling the bewildering scales of nonhuman systems into "excessive" narrative conventions.[8]

In the context of the late nineteenth century, these conventions were linked with specific genres considered oppositional to realism. Yet much of our received understanding of their excessive quality can also be traced to late-Victorian disputes over fiction's aesthetic values and social functions—debates that served to codify these generic distinctions. This essay will demonstrate that Hardy's position in what Jed Esty calls the late-Victorian "Realism Wars" was shaped by the formal challenge of representing the unprecedented scales of emerging scientific knowledge.[9] Ironically, Hardy's most direct engagements with that knowledge seem to gravitate toward the very forms that had been ostensibly surmounted by the realist novel's "scientific methods."[10] Recognizing the ways in which Hardy's "excessive" moments serve narrative ends, I will suggest that we should regard them less as sensationalized (or generic, in a pejorative sense) exaggerations of reality, and more as strategic attempts to represent reality beyond realism. Hardy's novels thus turn a scalar problem into an opportunity: By framing scenes of human life within scales that render them "infinitesimal," Hardy extends the novel's range of sympathy to subjects far beyond the human.

Scale, Realism, and Romance

My suggestion that excessive scales of scientific knowledge impose not just ontological challenges for human beings, but representational challenges for narrative forms structurally modeled on them, depends on recognizing scalar properties intrinsic to the novel itself.[11] These have been explored at length by Mark McGurl, who has recently argued that the "rise and subsequent history of the novel" might be characterized as a gradual "compression" of narrative form, whereby the novel overcomes a "problem of scale" by focusing narrowly on the human, and excluding what lies beyond it:

> Whether pitched at the level of small-scale intimacies or straining toward a grasp of the entire social system, the limits of the novel are defined by the limits of the human—which, to be sure, leaves space enough for a discourse of majestic complexity.[12]

McGurl's argument corresponds with theories of the novel that regard human limitations as productive constraints. These spatio-temporal structures are elaborated in Bakhtin's influential work on the chronotope,[13] though they are perhaps best exemplified by Lukács's concept of "biographical form." What makes the novel distinctive as a modern narrative system, Lukács contends, is its "refusal of the immanence of meaning to enter into empirical life"—that is, its commitment to depicting a demystified world, apprehended through verifiable observation. This, however, "produces a problem of form" because liberating the novel from immanent meaning also means shedding prescribed narrative *telos*, exposing it to what Lukács calls "a 'bad' infinity."[14] The novel "therefore needs certain imposed limits in order to become form," limits it acquires from the pattern of an individual's life story, whereby a "heterogeneous mass of isolated persons, non-sensuous structures and meaningless events receives a unified articulation by the relating of each separate element to the central character and the problem symbolized by the story of his life."[15] Biographical form supplies narrative structure while nevertheless (as Lukács clearly understands) shifting a tremendous amount of symbolic meaning onto individuals, registering the significance of all external events in terms of their character development. As Ian Watt explains in *The Rise of the Novel*, the "novel's closeness to the texture of daily experience directly depends upon its employment of a much more minutely discriminated time-scale than had previously been employed in narrative."[16] Narrative significance, in this account, again depends on adhering to scales that meaningfully frame the lifespan of the individual subject.

Yet these two central commitments—"closeness to the texture of daily experience" and fidelity to empirically observed reality—seem to enter into contradiction in the late Victorian period. How, for example, could the temporality of daily life be reconciled with the timescale of the geological epoch or, indeed, emerging theories of entropy that anticipated the extinction of the sun, followed eventually by the "heat death" of the universe, "the end of all physical phenomena"?[17] George Levine suggests that under these conditions "the realist exploration of reality becomes a necessary and self-destructive act. To attain knowledge is to achieve integrity and stature at the expense of finding oneself the butt of the great cosmic joke."[18] The expanding power of scientific observation during this period, moreover, contributed to a sense of spatial dislocation that accompanied the late-century globalization of Western financial and political networks. For the individual, writes Fredric Jameson, the "structural coordinates" of daily life were distorted to such a degree that they became "no longer accessible to immediate lived experience."[19] This in turn produced a contradictory "situation in which we can say that if individual experience is authentic, then it cannot be true; and that if a scientific or cognitive model of the same content is true, then it escapes individual experience."[20] Jameson and Levine agree that the narrative form most profoundly affected by this "crisis of representation" was the realist novel, whose reliance on "the phenomenological experience of the individual subject—traditionally, the supreme raw materials of the work of art—[became] limited to a tiny corner of the social world."[21] Yet while Jameson holds that the realist novel was drawn into a spatial cul-de-sac that later artists sought to escape through modernist innovations, Levine contends that "[a]ny fiction that confronts this late, disenchanted, and dualistic vision must, like much late-century fiction, take the shape of romance. It can make no final accommodation, in the realist tradition, to the culture it purports to describe or to the audience it addresses."[22]

Thus, rather than looking ahead to twentieth-century modernism, we might turn to the "romance revival" of the 1880s and 90s for evidence of an alternative, if reactionary, response to the same formal dilemma. The romance revival would then signal one way in which late-Victorian novelists were using genre to think in terms of narrative scale. The commercial success of works by H. Rider Haggard, Robert Louis Stevenson, Grant Allen, and H. G. Wells, among others, occasioned fierce disputes over the merits of romance, as well as its cultural implications. Andrew Lang famously weighed in on this debate in his 1887 essay "Realism and Romance," offering a pretext of neutrality before claiming that "the great heart of the people," which "demands tales of swashing blows, of distressed maidens rescued," could not be satisfied by an anemic fare of "accurate minute descriptions of life as its is lived, with its most sordid forms carefully elaborated."[23] Lang stages his defense of romance via a deftly feinted attack on realism, classifying it not as the absence of generic conventions—but as a genre in its own right:

> One only begins to object if it is asserted that this *genre* of fiction is the only permissible *genre*, that nothing else is of the nature of art.... Were I in a mood to disparage the modern Realists (whereas I have tried to show that their books are, in substance, about as good as possible, granting the *genre*), I might say that they not only use the microscope, and ply experiments, but ply them, too often, in *corpore vili*.[24]

Lang goes on to criticize "modern Realists" for "a sort of cruelty and coldness in their dealings with their own creations," and for a mannered interiority possessing "an almost unholy knowledge of the nature of women."[25] These ostensibly literary arguments mediate competing attitudes about how to scale Britain's popular imagination: whether outward, toward the horizons of an expansionist empire, or inward, toward subjects confined within national borders. "Realism wars," Esty argues, "name the fundamental ideological clash between romantic utopianism (shining universal values without territorial limits) and realist authority (representation of social conditions as delimited by societal space and historical time)."[26] Lang's objections to realism's cloistered interiority are motivated by his desire for spatial escalation, for "imaginations to fire in a more heated and more worldly direction, globalizing and vitalizing the literary culture."[27]

Hardy also participated in the late-Victorian realism wars, though his allegiance varied at different stages in his career. Hardy wrote several "Romances"—including *A Pair of Blue Eyes* (1873) and *Two on a Tower* (1882)—which are perhaps more attuned to "the nature of women" than to the style of adolescent adventure that Lang credits as roborative to "the great heart of the people." Nevertheless, they are similarly marked, as I will discuss shortly, by attempts to move beyond the scale of the realist novel. The fact that these works occupy a marginal status in Hardy's oeuvre is largely a result of their author's deliberate attempts to influence their critical reception. Hardy used the "Wessex Edition" (1912–31) as an opportunity both to make extensive revisions to his novels and, just as significantly, to classify them into three categories: Novels of Character and Environment, Romances and Fantasies, and Novels of Ingenuity. Descending in order from most to least realistic—and in Hardy's estimation at the time, from most to least artistically accomplished—these classifications helped to formalize a late-career investment in the author's realist credentials. As the title of the "Wessex Edition" suggests, Hardy laid claim to those credentials through unity of place, even though this unity was achieved largely after the fact.[28] Hardy's 1912 General Preface to the Wessex Edition assures "keen hunters for the real," particularly "readers interested in landscape, prehistoric antiquities, and especially old English architecture, that the description of these backgrounds has been done from the real—that is to say, has something real for its basis, however illusively treated."[29] Yet Hardy's repetition of the word "real" seems to betray an anxiety that, in spite of their paratextual accuracies, his novels' *plots* might still seem sensational or even fantastic—and he remained acutely aware that his romances in particular had been criticized for excesses and exaggerations.[30] At the same time, he increasingly felt a personal responsibility to provide accuracy as the chronicler of a rural way of life on the brink of extinction:

> [I]f these country customs and vocations, obsolete and obsolescent, had been detailed wrongly, nobody would have discovered such errors to the end of Time. Yet I have instituted inquiries to correct tricks of memory, and have striven against temptations to exaggerate, in order to preserve for my own satisfaction a fairly true record of a vanishing life.[31]

Despite these avowals, Hardy's earlier opinions about the aesthetic merits of realism were far less conciliatory. For example, in his essay "The Science of Fiction" (1891) Hardy refutes what he calls "scientific realism," arguing that its proponents' obsession with an increasingly minute and detailed reporting of daily life could not produce a meaningful account of its subjects. Hardy grants that the demand from certain critics to make fiction ever more "scientific"—in the sense of observational accuracy—"appears to owe its origin to the just perception that with our widening knowledge of the universe and its forces, and man's position therein, narrative, to be artistically convincing, must adjust itself to the new alignment."[32] However, he rejects the notion that fiction is capable of such an adjustment. Narrative cannot achieve a scientific "copying" of reality, he claims, because of "the impossibility of reproducing in its entirety the phantasmagoria of experience with infinite and atomic truth, without shadow, relevancy, or subordination."[33] And even if it were hypothetically possible to reach a state of total objectivity,[34] the result would not be recognizable as a work of art: the "attempt to set forth the Science of Fiction in calculable pages is futility; it is to write a whole library of human philosophy, with instructions how to feel."[35] Hardy's *reductio ad absurdum* argument against scientific realism corresponds with his view, expressed elsewhere, that

> Art is a disproportioning—(i.e. distorting, throwing out of proportion)—of realities, to show more clearly the features that matter in

those realities which, if merely copied or reported inventorially, might possibly be observed, but which more probably would be overlooked. Hence "realism" is not Art.[36]

Rather than mimesis, then, Hardy was an early and vocal advocate of estrangement as an artwork's definitive, essential function. His romances, as I will discuss in the remainder of this essay, aim to throw reality "out of proportion" not by introducing unrealistic content, but rather by incorporating the deranging—but all too real—scales involved in Hardy's "widening knowledge of the universe and its forces."

The Transfer of Excess

Scale is the central problem of *Two on a Tower*, a novel whose action takes place mostly on a small country estate, but whose narrative reach extends across several continents and into the depths of outer space. Written in haste for serialization in *The Atlantic*, Hardy regarded it an ambitious failure. Its plot focuses on the relationship of Lady Viviette Constantine, a lonely upper-class woman in her late-twenties who has been abandoned by her abusive husband (adventuring in Africa, later presumed dead), and Swithin St. Cleeve, a handsome middle-class youth in his early twenties who fervently desires to make a name for himself as an astronomer. The two are brought together when Viviette discovers that Swithin has appropriated a large, isolated tower on her estate to record his observations of the stars. The ancient tower under the night sky becomes a richly symbolic setting for conjuring romantic associations, and as the strangers become secretly acquainted there, tropes of chivalry and courtly love are transposed into the pursuit of cosmic wonders and the scientific conquest of strange worlds.

Partly in acknowledgement of these themes, Hardy subtitled the novel "A Romance." But if this phrase was also, as Richard Nemesvari writes, "calculated to suggest a lowering of expectations," Hardy lowered them further in his 1895 preface by calling it a "slightly-built romance."[37] Yet in the same preface Hardy also claims that the novel aspires to the highest possible ambition: "to set the emotional history of two infinitesimal lives against the stupendous background of the stellar universe, and to impart to readers the sentiment that of these contrasting magnitudes the smaller might be the greater to them as men."[38] Critics have noted the implied contradiction of this statement, "that any fiction appropriately defined as a 'slightly-built romance' would be able to sustain such a grandiose thematic burden."³⁹ However, the apparent disparity between form and content is, perhaps, appropriate for a novel whose motivation is disproportion itself, a theme that is reiterated by the relative (and problematic) differences in age, experience, class, and marital status between Swithin and Viviette.

Studying astronomy on the tower becomes a strategy for social leveling: Victorian mores are minimized by the novel's sustained contemplation of the cosmic universe, which makes the disparities between characters seem insignificant by comparison: "His vast and romantic endeavors lent him a personal force and charm which she could not but apprehend. In the presence of the immensities that his young mind had, as it were, brought down from above to hers, they became unconsciously equal."⁴⁰ In a letter to Edmund Gosse, Hardy explained his method: "I send this particular book in the belief that you will perceive, if nobody else does, what I have aimed at—to make science, not the mere padding of a romance, but the actual vehicle of romance."⁴¹ The generic form is enabled by, and informed by, astronomy itself. The novel's accurate descriptions of scientific instruments—telescopes, observatories, mathematical calculations—serve not as "mere padding" or realistic filler, but as devices for framing romance on a "universal" scale.

However, exposing the novel to the size of the cosmos introduces scale effects. While such extreme proportions serve to minimize differences between the protagonists, they also threaten to overwhelm the significance of the everyday actions and events that convey narrative significance, rendering the plot irrelevant. We encounter this problem early on when Swithin, educating Viviette in the basics of astronomy, attempts to relate to her the scale of the universe. Viviette has actually come to the tower to discuss "a personal matter," but after Swithin begins to explain his topic she perceives hers to be unimportant by comparison. "Let us finish this subject first; it dwarfs mine," she says. "Thereupon [Swithin] took exception to her use of the word 'grand' as descriptive of the actual universe."⁴² Swithin patronizingly cautions Viviette against dispelling her ignorance of astronomy, for once she glimpses the actual scale of the cosmos, she will experience a displacement so extreme that it will "dwarf" not just "personal matters," but the descriptive economy of language itself. This warning only serves to whet Viviette's appetite and Swithin, eager to impress her, "tried to give her yet another idea of the size of the universe":

> "There is a size at which dignity begins," he exclaimed; "further on there is a size at which grandeur begins; further on there is a size at which solemnity begins; further on, a size at which awfulness begins;

further on, a size at which ghastliness begins. That size faintly approaches the size of the stellar universe. So am I not right in saying that those minds who exert their imaginative powers to bury themselves in the depths of that universe merely strain their faculties to gain a new horror?"[43]

Swithin's speech treats cosmic scale as a discursive problem; despite his technical expertise, he cannot relate "the immeasurable" except by "figures of speech," "apt comparisons," and "leading-strings." Hardy here puns on the concept of "figure," shifting its connotations from the mathematical to the rhetorical. The "twenty million" stars that are visible to Swithin's telescope are insignificant in relation to the unfathomable voids that Swithin calls "Impersonal monsters, namely, Immensities":

> Until a person has thought out the stars and their inter-spaces, he has hardly learnt that there are things much more terrible than monsters of shape, namely, monsters of magnitude without known shape. Such monsters are the voids and waste places of the sky.... Those deep wells for the human mind to let itself down into, leave alone the human body![44]

"The body," Susan Stewart explains, "is our mode of perceiving scale," and "the body of the other becomes our antithetical mode of stating conventions of symmetry and balance on the one hand, and the grotesque and disproportionate on the other."[45] Swithin's monstrous figures threaten Swithin's embodied *sense* of scale: ascending from "dignity" and "grandeur" to "solemnity... awfulness... ghastliness... horror," Swithin describes a scale of affects produced by larger and larger cosmic bodies in relation to the human body—a range reminiscent of what Fredric Jameson calls the "chromaticism of the body itself."[46] Viviette, too, confronts the monstrous size of the universe as an embodied experience, crying out: "Oh, pray don't; it overpowers me!... It makes me feel that it is not worth while to live. It quite annihilates me!"[47]

Viviette's fear of annihilation also indicates why *Two on a Tower*'s cosmic scales are formally dangerous: they have the effect of obliterating its human plot. When Swithin attempts to steer Viviette back to the "personal matter" she came to discuss, she feels unable to proceed with matters that now seem trivial:

> "The immensity of the subject you have engaged me on has completely crushed my subject out of me! Yours is celestial; mine lamentably human! And the less must give way to the greater."

"But is it, in a human sense, and apart from macrocosmic magnitudes, important?" . . .

"It is as important as personal troubles usually are."[48]

Swithin and Viviette's dialogue effectively dramatizes the novel's formal dilemma: How can the lives of characters who "feel human insignificance too plainly," who realize that "nothing is made for man," be made to seem "important"?[49] This disproportioning effect is not unidirectional, however, because the distance between cosmic events and those on Earth makes the former seem just as insignificant as the latter:

"What do you see?—something happening somewhere?"

"Yes, quite a catastrophe!" he automatically murmured, without moving round.

"What?"

"A cyclone in the sun."

The lady paused, as if to consider the doubtful weight of that event in the scale of terrene life.

"Will it make any difference to us here?" she asked. . . .

"Ah, no."[50]

Shifting between cosmic events and the "scale of terrene life," the novel fluctuates between the domestic, the personal, and the everyday—where it cultivates a mode associated with realism—and "immensities" whose "powers are so enormous, and weird, and fantastical" that they cause the novel itself to shift into another generic mode.[51] Thus, while Hardy's decision to label *Two on a Tower* a "romance" may signal an awareness that the cosmic ultimately gets the better of the sublunary, his foregrounding of the work's generic status also suggests that the sheer scale of astronomical observations, empirically accurate in themselves, introduces a mode of excess that makes the narrative nonrealistic.

To accommodate the excessive figures of cosmic monstrosities, Hardy heightens the domestic plot through what could be described as a transfer of excess, whereby the "scale of terrene life" is endowed with some of their "enormous, and weird, and fantastical" powers. When, for example, Swithin is reunited with an older Viviette, he perceives that "another woman sat before him, and not the original Viviette. . . . the masses of hair that were once darkness visible had become touched here and there by a faint grey haze, like the Via Lactea in a midnight sky."⁵² Hardy takes what is already a baroque simile—greying hair resembling the Milky Way—and exaggerates it by translating it into Latin, a commentary perhaps on Swithin's inability to turn his mind away from his true desire, astronomy, and its professional nomenclature. We might identify its broader strategic value by reading this Miltonic description ("darkness visible") as a melodramatic one. "Excess," as Peter Brooks influentially contends, "is in fact inherent to the form" of melodrama; it

> is ever implicitly an emblem of the cosmic ethical drama, which by reflection illuminates life here below, makes it exciting, raises its stakes. Hence melodrama's mode must be centrally, radically hyperbolic, the mode of the bigger-than-life, reaching in grandiose reference to a noumenal realm.⁵³

But Brooks's identification of the "play of cosmic moral relations and forces" is complicated in this case by the fact that the novel places the cosmic at its center, not just reaching out to it referentially but channeling its grandiosity directly into the quotidian, making the commonplace simultaneously lavish and precarious.⁵⁴ The plot of *Two on a Tower* is exaggerated by marital scandal, secret histories, sadistic villains, legal documents riddled with disastrous conditions, accidents, overheard conversations, misplaced letters, mistaken identity, unplanned parenthood, and somnambulism. Indeed, on the last page, when "[s]udden joy after despair" wrenches her "overstrained heart too smartly," Viviette's heart literally breaks.⁵⁵ Such extravagant, even scandalous, elements might also qualify *Two on a Tower* as a "sensation novel," "a genre whose shock-appeal," writes Emily Steinlight, depends "on the disclosure of a calamitous surplus at the heart of the domestic sphere."⁵⁶ By a process of transfer or reversal, the domestic sphere in *Two on a Tower* is tasked with supporting the "calamitous surplus" of the astronomical theme, a theme it attempts to accommodate by raising the stakes of the everyday to hyperbolic proportions. Thus the excessive mode associated with melodrama and romance seems here to provide a certain formal affordance, one that operates across Hardy's fiction—the capacity to narrate scales beyond the human.

Deep Suspense

Some property or characteristic of the inhuman seems to permeate the especially bleak outcomes of Hardy's novels. George Levine observes that Hardy's "extravagance" in *Two on a Tower* seems "almost to be parodying his later cosmically pessimistic self."[57] Indeed, this morose attitude has been the frequent subject of reviews and critical studies, such as Forsyth's "The Pessimism of Mr. Thomas Hardy" (1912), which maintains that Hardy's "mostly crushing" attempts to frame "man's universal and final question" on the "scale of the world" make each of us "objects of pity, and our best social ethic rests on a proper pity for our fellow-victims."[58] Ernest Brenneke's *Thomas Hardy's Universe* (1924) identifies the influence of the archpessimist Schopenhauer, who notoriously asserted that "the enormous amount of pain that abounds everywhere in the world, and originates in needs and necessities inseparable from life itself," serves "no purpose at all and [is] the result of mere chance."[59] "Cosmic pessimism" is so overwhelming, according to Eugene Thacker, that it dissolves pessimism itself as a philosophical position:

> The contours of cosmic pessimism are a drastic scaling-up or scaling-down of the human point of view, the unhuman orientation of deep space and deep time, and all of this shadowed by an impasse, a primordial insignificance, the impossibility of ever adequately accounting for one's relationship to thought—all that remains of pessimism is the desiderata of affects—agonistic, impassive, defiant, reclusive, filled with sorrow and flailing at that architectonic chess match called philosophy, a flailing that pessimism tries to raise to the level of an art form (though what usually results is slapstick).[60]

It is difficult not to acknowledge the similarity between this outlook and the notebook entries described at the beginning of this essay, in which Hardy doubts whether the "materials for happiness to higher existences" can be found on this or any planet.

One could speculate that Hardy might have spared his characters—fictional representations of human beings who, by virtue of self-awareness, might even constitute a distinct form of "higher existence"—much misfortune if he had taken his own proposition differently. By imagining favorable conditions for higher existences on other planets, he might have conceived the form of utopian scientific romance that was taking shape at almost the very moment that he, having allegedly scandalized the public, stopped writing novels. Instead, Hardy remained committed to building worlds that closely resembled his own, and while his "Romances" use scientific themes to gesture toward alternative worlds in the deep past or on the cosmic horizon, they ultimately refuse the possibility of escaping the material realities of *this* planet.

Nevertheless a sense of hopefulness also emerges from such extreme perspectives. Hardy's novels generate strange forms of sympathy by communicating across distant times and places, between the real and the unreal, between the human and the nonhuman. Nowhere is this more evident than in *A Pair of Blue Eyes*, in the famous scene in which a geologist named Henry Knight finds himself dangling from a cliff over the sea, clinging on for dear life. From this precarious position he chances to see a fossilized trilobite embedded in the rock:

> Separated by millions of years in their lives, Knight and this underling seemed to have met in their death. . . . [A]t this dreadful juncture his mind found time to take in, by a momentary sweep, the varied scenes that had had their day between this creature's epoch and his own. There is no place like a cleft landscape for bringing home such imaginings as these.
>
> Time closed up like a fan before him. He saw himself at one extremity of the years, face to face with the beginning and all the intermediate centuries simultaneously . . . the lifetime scenes of the fossil confronting him were a present and modern condition of things.[61]

Knight sees not his own life, but all previous forms of life on earth pass before his eyes. The narrator's free-indirect observation concerning the "present and modern conditions of things" involves imagining the simultaneous presence of the entire fossil record, and it is surprising that such a sobering theme is not recollected in a moment of thoughtful tranquility, but at the peak of a desperate action scene. "Finding time" to "take in" these perspicacious thoughts jars with the immediacy of Knight's personal struggle for existence, but in suspending the fall, Hardy is also drawing out the scene's *suspense*.

Knight, whose name carries obvious connotations of romance, seems larger than life in this moment because his predicament is vastly excessive to the kinds of events that typify daily life. While it is fair to suggest that the scene is "unrealistic" insofar as it relies on an extraordinary, improbable circumstance to dramatize humanity's relationship to geological time, this criticism becomes a negative one only if the novel's aesthetic and moral value ultimately depends on its strict adherence to realistic conventions. Knight's momentary encounter with the trilobite is "separated by millions of years," a distance that both complicates and reinforces their relationship as individuals. Shifting into a heightened mode, Knight's individual extinction becomes symbolic of the human species as a whole, which must take its place alongside even the lowest invertebrates which it will "meet in death." Given Hardy's views on the evolutionary misstep of "crossing the line from invertebrates to vertebrates," it is plausible that Knight might even regard this "underling" with a degree of envy. Yet if the trilobite represents more than an elaborate *memento mori*, it is because the scene asks its reader to grasp a much wider view of species interconnectedness as an urgent matter of survival. What this scene ultimately gives us by abandoning the limits of realism is a vision of deep time simultaneous with the present and a sensational narrative tableau—a device so evocative that it would come to be called "the cliffhanger."

Notes

1. Quoted in Florence Emily Hardy, *The Early Life of Thomas Hardy, 1840–1891* (Cambridge: Cambridge University Press, 2011), 285–86.

2. Cf. Hardy's 1909 letter: "The discovery of the law of evolution, which revealed that all organic creatures are of one family, shifted the centre of altruism from humanity to the whole conscious world collectively." Quoted in Florence Emily Hardy, *The Later Years of Thomas Hardy, 1892–1928* (New York: Macmillan, 1930), 138. For a sustained examination of Hardy's positions on consanguinity and ethics, see Elisha Cohn, *Still Life: Suspended Development in the Victorian Novel* (New York: Oxford University Press, 2016).

3. The quotidian form of the diary itself serves to underscore this disparity: Hardy's entries begin by recording the rhythm of daily routine ("May 5. Morning. Sunday. To Bow Church, Cheapside, with Em . . .") but often end with a shift of perspective that calls into question the significance of events perceived on a human scale (" . . . That which, socially, is a great tragedy, may be in Nature no alarming circumstance."). *Early Life*, 286.

4. Hardy's work, according to Gossin, is always overshadowed by this overwhelming question: "how can anyone cope with the truly frightening fatal flaw in nature's unintended consequences, namely, that beneath an *astronomical* sky, filled with evolving and decaying stars, upon this actively *geological* planet, *biological* beings should have ever evolved *consciousness* of our own evolution and mortality?" Pamela Gossin, *Thomas Hardy's Novel Universe: Astronomy, Cosmology, and Gender in the Post-Darwinian World* (Aldershot: Ashgate, 2007), 116–17.

5. Gillian Beer, *Darwin's Plots: Evolutionary Narrative in Darwin, George Eliot and Nineteenth-Century Fiction*, 3rd ed. (Cambridge: Cambridge University Press, 2009), 233.

6. See Timothy Clark, "Scale," in *Telemorphosis: Theory in the Era of Climate Change*, ed. Tom Cohen, vol. 1 (Ann Arbor: Open Humanities Press, 2012), 148–66.

7. Even formal structures of "rational" thought such as ideas and arguments implicitly depend on "ratios," such that, as Clark notes, "what is self-evident or rational at one scale may well be destructive or unjust at another." Ibid., 150.

8. Hardy, paraphrasing Wordsworth, argued that "a more natural magic" than realism was needed to express the "still sad music of humanity." *Thomas Hardy's Personal Writings: Prefaces, Literary Opinions, Reminiscences*, ed. Harold Orel (Basingstoke: Macmillan, 1990), 135. On reenchantment and scientific consciousness, see Morris Berman, *The Reenchantment of the World* (Ithaca: Cornell University Press, 1981). See also, George Lewis Levine, *Darwin Loves You: Natural Selection and the Re-Enchantment of the World* (Princeton: Princeton University Press, 2008).

9. Jed Esty, "Realism Wars," *NOVEL: A Forum on Fiction* 49, no. 2 (August 2016): 316–42.

10. The association of scientific methods with the realist novel's techniques is, of course, a critical convention that was established first by the late-Victorians themselves. Stephen Arata offers the example of Mrs. Humphrey Ward, who in 1884 expressed "the widely accepted truth that 'the French have now fully possessed themselves of those realistic and "scientific" methods, which are only just now beginning to affect the English novelist.'" Stephen Arata, "Realism," in *The Cambridge Companion to the Fin de Siècle*, ed. Gail Marshall (Cambridge: Cambridge University Press, 2007), 172.

11. For a helpful account of scale and the development of the novel, see Elizabeth Deeds Ermarth, *Realism and Consensus in the English Novel: Time, Space and Narrative* (Edinburgh: Edinburgh University Press, 1998).

12. Mark McGurl, "Gigantic Realism: The Rise of the Novel and the Comedy of Scale," *Critical Inquiry* 43 (Winter 2017): 18.

13. M. M. Bakhtin, *The Dialogic Imagination: Four Essays*, ed. Michael Holquist, trans. Caryl Emerson and Michael Holquist (Austin: University of Texas Press, 1981).

14. György Lukács, *The Theory of the Novel: A Historico-Philosophical Essay on the Forms of Great Epic Literature* (London: Merlin Press, 1971), 71, 81.

15. Ibid., 81. See also Jed Esty, *Unseasonable Youth: Modernism, Colonialism, and the Fiction of Development* (New York: Oxford University Press, 2012), 27.

16. Ian Watt, *The Rise of the Novel: Studies in Defoe, Richardson and Fielding* (Berkeley: University of California Press, 1957), 21.

17. W. J. Macquorn Rankine, quoted in Martin Meisel, *Chaos Imagined: Literature, Art, Science* (New York: Columbia University Press, 2015), 385. See also George Lewis Levine, *Dying to Know: Scientific Epistemology and Narrative in Victorian England* (Chicago: University of Chicago Press, 2002), 201.

18. George Lewis Levine, *The Realistic Imagination: English Fiction from Frankenstein to Lady Chatterley* (Chicago: University of Chicago Press, 1981), 53.

19. Fredric Jameson, "Cognitive Mapping," in *Marxism and the Interpretation of Culture*, ed. Cary Nelson and Lawrence Grossberg (Urbana: University of Illinois Press, 1988), 349.

20. Ibid.

21. Ibid.

22. Levine, *Realistic Imagination*, 53.

23. Andrew Lang, "Realism and Romance," *Contemporary Review* 52 (1887): 684–85.

24. Ibid., 687, original emphasis.

25. Ibid., 688.

26. Esty, "Realism Wars," 328.

27. Ibid., 331.

28. As Richard Nemesvari points out, "Hardy only gradually came to understand its potential as a symbolic landscape and realist device." *Thomas Hardy, Sensationalism, and the Melodramatic Mode* (New York: Palgrave Macmillan, 2011), 18.

29. Hardy, *Personal Writings*, 47, 46.

30. Recalling the initial reception of *Two on A Tower*, Hardy wrote, "I was made to suffer in consequence of several eminent pens, such warm epithets as 'hazardous,' 'repulsive,' 'little short of revolting,' 'a studied and gratuitous insult,' being flung at the precarious volumes." *Personal Writings*, 16–17.

31. Ibid., 46.

32. Ibid., 135.

33. Ibid.

34. For an account of "The Science of Fiction" as Hardy's call for an "impassioned" objectivity, see S. Pearl Brilmyer, "Impassioned Objectivity: Nietzsche, Hardy, and the Science of Fiction," *Boundary 2* (October 5, 2016), http://www.boundary2.org/2016/10/s-pearl-brilmyer-impassioned-objectivity-nietzsche-hardy-and-the-science-of-fiction/.

35. Hardy, *Personal Writings*, 138.

36. Thomas Hardy, *The Life and Work of Thomas Hardy*, ed. Michael Millgate (London: Macmillan, 1984), 239.

37. The preface responds to what were mostly tepid reviews of the novel, and Hardy was at that point eager to establish distance from it. Hardy, *Personal Writings*, 16; Nemesvari, *Thomas Hardy*, 20.

38. Hardy, *Personal Writings*, 16.

39. Simon Gatrell, *Hardy, the Creator: A Textual Biography* (Oxford: Oxford University Press, 1988), 188.

40. Thomas Hardy, *Two on a Tower: A Romance*, ed. Sally Shuttleworth (New York: Penguin, 1999), 31.

41. Quoted in Sally Shuttleworth, Introduction to Thomas Hardy, *Two on a Tower: A Romance* (New York: Penguin, 1999), xvii.

42. Hardy, *Two on a Tower*, 29.

43. Ibid., 30.

44. Ibid.

45. Susan Stewart, *On Longing: Narratives of the Miniature, the Gigantic, the Souvenir, the Collection* (Durham, N.C.: Duke University Press, 1993), xii.

46. Fredric Jameson, *The Antinomies of Realism* (London; New York: Verso, 2015), 42.

47. Ibid., 29.

48. Ibid., 31.

49. Ibid., 28.

50. Ibid., 8.

51. Ibid., 57.

52. Ibid., 259.

53. Peter Brooks, *The Melodramatic Imagination: Balzac, Henry James, Melodrama and the Mode of Excess* (New Haven: Yale University Press, 1996), 54.

54. Ibid.

55. Hardy, *Two on a Tower*, 262.

56. Emily Steinlight, "Why Novels Are Redundant: Sensation Fiction and the Overpopulation of Literature," *ELH* 79, no. 2 (2012): 502.

57. Levine, *Dying to Know*, 28.

58. P. T. Forsyth, "The Pessimism of Mr. Thomas Hardy," *The Living Age* (November 23, 1912), 480.

59. Ernest Brennecke, *Thomas Hardy's Universe: A Study of a Poet's Mind* (New York: Russell & Russell, 1966); Arthur Schopenhauer, "On the Sufferings of the World," in *Studies in Pessimism*, trans. T. Bailey Saunders (New York: Wiley, 1942).

60. Eugene Thacker, "Cosmic Pessimism," *Continent* 2, no. 2 (January 8, 2012): 68, http://www.continentcontinent.cc/index.php/continent/article/view/84.

61. Thomas Hardy, *A Pair of Blue Eyes*, ed. Pamela Dalziel (New York: Penguin, 1998), 213–14.

PART IV
Futures

CHAPTER 10

Electric Dialectics
Delany's Atlantic Materialism
Monique Allewaert

> A commodity appears at first sight an extremely obvious, trivial thing. But its analysis brings out that it is a very strange thing abounding in metaphysical subtleties and theological niceties. . . . The form of wood, for instance, is altered if a table is made out of it. Nevertheless the table continues to be wood, an ordinary sensuous thing. But at soon as it emerges as a commodity, it changes into a thing which transcends sensuousness. It not only stands with its feet on the ground, but, in relation to all other commodities, it stands on its head, and evolves out of its wooden brain grotesque ideas, far more wonderful than if it were to begin dancing of its own free will.
>
> —KARL MARX, *Capital*, vol. 1

Spiritualism in the Key of Materialism

Marx's chapter, "The Fetishism of the Commodity and Its Secret," famously opens with the figure of a table on its feet turned upside down, in which position it spins grotesque fantasies. Right side up, the table's relation to its sensual physicality and its usefulness in the household or shop are paramount. Upside down, its only characteristic is a phantasmagoric exchange value. For Marx, this vignette crystallizes the central dynamic of theory under industrial capitalism: It depends on a dualistic logic that divides out what it counts as material from what it counts as ideal and then prioritizes the ideal. This ends up cutting human beings off from the materialism (the dialectical process through which phenomena, milieu, and history emerge) that Marx posits as its first principle and as the key to any political and social liberation. If this vignette condenses Marx's critique of theory under capitalism, what's scarcely been noted is that it culminates in a relative clause that references tables that are neither right side up nor upside down but "dancing of [their] own free will." To explicate this clause, Marx adds in a footnote that "China and the tables began to dance when the rest of the world appeared to be standing still—*pour encourager les autres*."[1] This note, as well as the translator's explication of it, indicates that these dancing tables reference the eruption of table-rapping spiritualisms.[2] These spiritualisms arrived on the scene in 1848 in upstate New York, launching a spiritualist craze that spread throughout the Americas and Western Europe over the 1850s.

Marx's reference to dancing tables seems a passing bit of silliness entirely typical of the ludic style that pulses through his economic analysis. Yet given *Capital*'s other more sustained engagements with American topics ranging from Robinson Crusoe stories to the centrality of colonies and slaves to primitive accumulation, we might link this joke to a more serious problem of what to make of the spiritualisms at stake in these dancing tables. If the commodity capitalism that Marx describes in *Capital* works by dividing out materiality from ideality, the tables dancing in the Americas are products of subcultures that, far from dividing materiality from ideality, insist on the emphatic materiality of spirituality itself, in so doing seizing "spiritual life" from the realm of the ideal—whether the idealism of conventional Christianity or that of the commodity capitalist. After all, these spirits were sensational precisely because they acted on and through sensuous phenomena like tables and set in motion processes—from allowing women to perform as public speakers to making the body the medium for and manifestation of spirituality—that changed social organizations from the family to the church. Taking the Americas' spiritualized materialisms seriously requires reading Marx's joke straight and taking what he relegates to the relative as the starting point of analysis. Yet this move also stays in the line of Marx. For he does not quite cast the American spiritualisms he references pejoratively. To the contrary, they encourage Europe's economists and laborers in a way not entirely unlike that which *Capital* itself aims to achieve.[3]

The dancing tables of Marx's joke came to international attention in 1849 when the Fox Sisters moved from a small New York hamlet to the booming western city Rochester, where they claimed to be able to channel spirit forces that manifested in the table rappings and tiltings that Marx teasingly calls dancing.[4] The efflorescence of spiritualized materialisms in Rochester was closely allied with the abolitionism also flourishing in that city. The Rochester-based abolitionist and women's rights activist Amy Post worked to integrate spiritualism and antislavery programs, beseeching Frederick Douglass and his paper to support the movement. Another white abolitionist, Harriet Beecher Stowe, was a proponent of mesmerism who joined the spiritualist movement. Stowe cast African Americans as having inherited that "peculiar temperament" that made them adept sorcerers, by which she meant fetishists, as well as powerful mesmerists.[5] Black abolitionists were more circumspect in their estimations of these spiritualisms circulating through Rochester. Frederick Douglass and Martin Delany's newspaper *The North-Star*, also headquartered in Rochester, reported on the spiritualism for which the city was becoming famous and Douglass attended at least one séance.[6] However, far more than their white American peers, black abolitionists emphasized that spiritualism must be understood as a materialism of this world that couldn't be confused with idealism, and they were chary to articulate their own positions on this spectacular materialism. Their caution was likely linked to the fact that blacks in the diaspora were themselves consigned to the realm of materiality—conceived quite literally as raw materials—and closed out of the culturally privileged position of ideality. The association of Africans in the diaspora with sorcery, fetishism, mesmerism, and spiritualism, as well as black public intellectuals' guarded response to these phenomena, recalls other kinds of fetishes at stake in the mid–nineteenth century: not the purely ideological commodity fetish at stake in Marx's upside-down-table but the actual and decidedly material fetishes that circulated in the Atlantic world and that return to the foreground when we draw out from Marx's joke the Atlantic history and geopolitics it evokes.

Drawing on this close association of American spiritualisms with Afro-American practices classed as fetishism, this article will explore Atlantic materialisms that require that we think with and through fetishisms. If this possibility emerges only obliquely in Marx's writing, this article attempts to unfold this possibility and to do so in ways that move within Marx's diagnosis yet pull from it a different planetary materialism than that which he advocates. Unlike the commodity fetish, which expresses the idealism of a fundamentally dualistic capitalistic culture, the Atlantic fetish is an expression of materialisms emerging from the conflux of Atlantic cultures and tends toward pluralistic or relational logics in which antagonisms and alliances emerge through the play of circumstances instead of being absolute or determined in advance.

To develop this argument, this article moves from metropolitan centers such as London, where Marx wrote *Capital*, to boomtowns such as Rochester, to investigate colonial frontier spaces of the eighteenth century where Atlantic fetishisms proliferated. If Atlantic fetishes were brutally repressed by mercantilist and then capitalist colonial powers, they did not disappear during the colonial period or the slow decline of slave-fueled, plantation-based colonialism in the nineteenth century. Instead, they migrated into and influenced American-style spiritualist materialisms that circulated on the frontiers of capital and in its centers. The second part of this article proposes that this migration of fetishism into spiritualist materialisms is particularly evident in Martin Delany's post *North-Star* development of electrical theories as well as in his application of these electrical theories in the quasi-fictional work *Blake* (1859–61). At stake in this analysis of Afro-American fetishes and the spiritualist electrical theories into which they migrated is identifying nineteenth-century alternatives to capitalistic divisions of nature and culture and modes of valuation that Marx's reference makes clear have always been present. Yet they have remained obscure when critics from Marx forward have interpreted the nineteenth century primarily from the perspective of the metropolitan centers and in terms of the forward-looking modernity augured by capital. We might access and resuscitate this Atlantic materialism via a historical materialism that develops not only from the agons particular to centers of power but also at its frontiers. From the perspective of metropolitan centers, these materialisms emerging from frontiers have seemed backward and backward-looking. Yet this backwardness is not a liability: Rather, it comes from and occasions a historical materialism whose power derives from its orientation toward the past, not the future.

Fetishism and Materialisms of the Atlantic World

Jason Moore's analysis of capitalism focuses on the frontiers of the European powers associated with the rise of capitalism, whether the transoceanic slave trade or the Baltic forests that were felled for ships that contributed to the emergence of a global market.[7] This orientation makes clear that capitalism has developed through the appropriation of cheap natures (labor, raw materials, energy, food) from these frontiers. This in turn makes colonial frontiers—often thought of simply in terms of a primitive accumulation thought to precede capitalism—as fully central to the development of capitalism as the expropriation that is more commonly the focus of Marxist analysis. Following Moore by attending to frontier spaces, yet considering them not from the vantage of the metropole but with attention to the histories and modes of knowledge emerging in them, brings to the fore something in excess of the cheap natures that Moore argues fuel capitalism. We also see nature thought not as inconsequential or cheap but as of great value because, among other things, it is recognized as a force that moves through and produces culture.

After the "discovery" of the Americas, the expansion of trade outside of Europe and the near and far East and into Sub-Saharan Africa and the Americas brought into contact regions that had hitherto not been in sustained contact. While post-feudal Christian Europeans as well as their near and far Eastern trading partners broadly agreed about what counted as riches, Spanish traders who valued gold and cotton were dumbfounded that the Amerindians they encountered valued "glasses, bottles, and jars" and other "trifles."[8] Similarly, what counted as riches to colonial merchants did not count as riches to their African trading partners, as the Dutch merchant William Bosman makes clear in recounting that when the Axim village on the Guinea Coast shot gold instead of bullets from their guns, British traders took this as a sign that the Axim wanted to initiate trade only to be blown to bits when they arrived for the exchange.[9] This instability in the valuation of riches that followed on the rise of truly global trade was central to fetishisms in the Atlantic world. As cultural theorist William Pietz has argued, Atlantic fetishes proliferated in colonial contact zones struggling to translate and transvalue the cultural codes of feudal Christian lineage, African lineage, Amerindian lineage, as well as those of emerging merchant capitalism.[10] Atlantic fetishes, then, emerge from a context in which there is not one world and with it a uniform system of valuation but in which many systems of valuation were overlaid.

Let's consider a particular case in order to understand what fetishism signifies in the Atlantic sense of that term. In 1758, the African-born slave-turned-maroon François Makandal was executed in Cap Français, St. Domingue (now Cap Haitien, Haiti) for producing fetish artifacts that the colonial court denounced as not only without value but also as metaphysical and physical poisons. Sébastien-Jacques Courtin, the colonial administrator who presided over the trail of François Makandal and others accused of fetishism in 1757–58, produced a "Mémoire" that describes these artifacts in detail. "Here is how one makes a [fetish]," Courtin writes, and then offers a recipe.[11] First one gathers a range of materials including bones, nails, roots of banana trees and cursed fig trees, holy water, incense, and the Catholic host. These ingredients are then put into a cloth with dirt and holy water and wrapped many times to form a packet. While binding together these ingredients, the fetish artisan makes invocations in Arabic, French, and Kreyol. The actuated artifact is then doused in holy water and sold.[12] These fetish artifacts integrate French and Christian symbols of great value to the French colonial state—from Catholic rituals to the coinage with which fetishes were sometimes sold—with what Courtin classed as having no value and diabolical significance. The mix was especially perturbing to French colonial authorities.

While Courtin is most attuned to fetishes' appropriation of Catholic symbols, his description gives evidence that African and perhaps Amerindian practices were also at stake in their production. The use of roots evinces an African tradition of herbal medicine that migrated into the Americas with the slave trade.[13] Courtin also describes the "feeding" of fetish artifacts as well as commissioners' adherence to dietary prescriptions after receiving them, which evokes West African practices of libation and taboos. The collection of fragments of bones might reference the mortuary practices of the Arawak, indigenous to St. Domingue yet nearly exterminated by the eighteenth century, who placed the bones of the dead in reliquary artifacts.[14] That fetishes emerged from diasporic Africans' efforts to negotiate different Atlantic cultures and their valuations is also evident in Courtin's attention to incantations from three different linguistic cultures used in their activation—French European (references to God and Christ in French), Sub-Saharan West African as well as North African (the Arabic chant "Allah Allah"), and American (Kreyol).

In addition to negotiating the diverse cultures of the Atlantic world, fetishes also negotiate their diverse natures. On this point, consider the semiotics presumed in their production and use. If, in Marx's diagnosis, capitalism operated by splitting materiality from ideality and privileging the latter, this was linked to and followed on a long-standing Christian tendency to divide out nature from culture. The Atlantic frontiers in which fetishes circulated didn't presume this division of nature from culture or materiality from ideality. Instead, they understood nature as the formative matrix for cultural expression, thus presuming that nature was expressed, in part if not in totality, through culture. This imbrication of what the Western European Catholics classed as "nature" in culture is evident in the mode of semiotics at stake in Atlantic fetish artifacts. Courtin's text makes clear that fetish artifacts are composed of partial objects. These partial objects function indexically. The terms *index* and *indexically* recur across art historian Suzanne Blier's analysis of the fragments integrated into post-contact African fetish artifacts.[15] Blier's use of the term corresponds with Charles Peirce's definition of indexes as the category of signs that "show something about things, by being physically connected with them."[16] This means that indexical signs cannot be parsed by the logic of linguistic and extralinguistic or nature and culture as, above all, what they indicate is a domain of language in which the materiality of the phenomenal world and the materiality of the sign are co-constitutive.

In short, the semiotics that undergird the production and use of fetishes—indexical signs in material relation with their surroundings—privileges an Afro-American conception of nature and its organization. To be sure, in the eighteenth century it was not only Africans in and out of the Americas who argued for the imbrication of nature and culture and of matter and spirit. One of the most important in Atlantic frontiers was mesmerism.[17] The practice arrived in St. Domingue in 1784, quickly becoming popular among white and mulatto populations who used the therapy on slaves as well as on themselves.[18] Mesmer's theory offered a fiercely materialist challenge to what he claimed were mystifying accounts of the movement of physical bodies. He argued for the existence of a "universally diffused fluid" by which what seemed to be standing still was, by virtue of this magnetic fluid, always in motion, albeit a motion that was imperceptible to human eyes yet manifest and that could be manipulated by the mesmerist's rituals.[19] On being imported into St. Domingue, mesmeric therapy was used to bring languishing slaves back to health so that they would be capable of work.[20] It may have been via this route that mesmerism was integrated into Atlantic fetishisms. Starting in the mid to late 1780s, slave owners in St. Domingue claimed that their slaves gathered in banana groves and other refuges on the edge of plantations to partake in rituals that fused fetishistic and mesmeric practices. Colonial courts responded by recirculating laws against fetishes passed in the wake of Makandal's execution just over twenty-five years earlier, treating mesmerism as practiced by Africans in the diaspora as an outgrowth of fetishism.[21] Despite the fact that blacks in the colony were legally prohibited from both fetishistic and mesmeric practices, both persisted and Mesmer later claimed that diasporic Africans' appropriation of mesmerism as one of the causes of the Haitian Revolution.[22]

Electric Dialectics

It was most obviously Anglo-Europeans who associated fetishism, mesmerism, and spiritualism. Yet writers in the black diaspora also explored the relation between them. Frederick Douglass, the most famous of all black abolitionists, repeatedly engaged these problems in his investigations of fetish roots across his autobiographical narratives and in vesting the black hero of *The Heroic Slave* (1852) with mesmeric power. Jared Hickman casts Douglass's fetishism as a key aspect of what Hickman calls a "black prometheanism" that departs from both Hegelian and Marxist theories of history by locating rebellion and freedom not in the realm of spirit or in that of materialism but in an immanent field without any absolute and in which gods act in this world.[23] While Douglass is the most famous black abolitionist to trace a line between fetishism, mesmerism, and spiritualism, his less-known, more itinerant co-editor and colleague, Martin Delany, offers a still more fully wrought Atlantic materialism that tracks the imbrication of material and spiritual phenomena on a single immanent plane.

Born to a free mother and an enslaved father, both first-generation creoles, Delany was partly self-educated and partly school educated, including a brief stint at Harvard before being expelled for being black.[24] After founding and editing a paper for Pittsburgh's free black community, he teamed up with Douglass to edit *The North-Star*, an association that lasted for two years but dissolved because of Delany's more radical positions on racial blackness (he asserted it as a positive value whereas Douglass did not), constitutionalism (he rejected the U.S. Constitution as a slave document whereas Douglass worked within its frame), and colonization (Delany came to believe that colonization might be the only way for blacks in the diaspora to avoid lives fully determined by anti-black racism).[25] At the very moment Marx was researching and writing *Capital* in London, Delany—in no one location for long—was writing and publishing his first and only known fiction, *Blake; or the Huts of America*, in serial form in the *Anglo-African Magazine* (1859–62). He was also producing articles on electrical theory for this magazine.[26] The *Anglo-African*'s editor, Robert Hamilton, placed the first of these articles, "The Attraction of Planets," just before the first published chapters of *Blake*: chapters 28, 29, and 30 (the chronologically first chapters of the work were published in later issues). These chapters concern modes of Atlantic knowledge ranging from Anglo-European technologies and epistemologies to conjure.

In the January 1859 article "The Attraction of Planets," and the February 1859 article "Comets," Delany theorizes electricity by overlaying secular materialist and theistic materialist cosmologies. On the one hand, he casts electricity in clearly physical terms, expressly appropriating atomistic theories. On the other, he supposes that electrical charges were passed on to material phenomena when planets and other bodies were "ushered out of almighty hands'" (18, 2), casting electricity as the charged residue of divine power continually acting in the physical universe. This formulation keeps open the possibility of a spiritual outside to the material world under investigation in his articles. Yet if Delany leaves open the possibility of a prime mover that exceeds the material world, his articles are less interested in this than they are in investigating the residue that follows on the touch between divine and physical phenomena. For this reason, his works echo the key premise of spiritualists also writing at this moment, namely that spiritualism might be thought in the realm of the material.

Theories of electricity were first substantively developed in the eighteenth century, including by Benjamin Franklin, whom Delany admired and whom he likely read to develop his own theories of electricity.[27] By the mid-1840s, publics in metropolitan centers, boomtowns, and colonial frontiers understood that scarcely sensible electrical impulses could be transmitted across enormous distances so that telegraphic communications could pass almost instantaneously. Spiritualists proposed that these same sorts of impulses could make communications traverse the seemingly impossible distance between the dead and the living, not via undersea cables but super-sensible human beings, often female and sometimes black, who were thought to manage and circulate electrical and magnetic charges such that impulses transmitted by the far away and the dead could be converted into messages to the nearby and the living.[28] Nineteenth-century publics associated the electromagnetic forces that scientists theorized with the spiritualist forces that radical reformers spoke of, in part because spiritualists used the language of electromagnetics and cast their experiments as sciences. When Samuel Morse attempted to gain funding from the U.S. Congress for a national telegraph system, the Congress suggested that if they funded the telegraph they would then have to fund all spiritualist experiments.[29]

Delany argues that electricity explains why things cohere in a system, whether the solar system or the animal body.[30] Ever since Newton's *Principia*, the correct answer to why planets and bodies move in systems had been gravitational force. It is almost certain that Delany would have known the then nearly ubiquitous Newtonian account of planetary motion. Indeed, in July of 1859 his fellow contributor to the *Anglo-African*, the Oberlin-educated George Vashon, published an account of astronomy in the *Anglo-African* that explicitly references Newton's gravitational theory.[31] Despite this familiarity, Delany chose to privilege an electrical account. Delany's production of idiosyncratic theories is probably linked to his itinerant education, which made him less committed to prevailing doctrines than more classically educated men like Vashon. Moreover, in the '50s when he wrote his electrical articles and *Blake*, he'd become particularly irritated with the clubbiness of academic communities and their sanctioned knowledges, which his experiences at Harvard made clear included racism.

It's likely that Delany understood his electrical account of the movement of planets was idiosyncratic in switching from gravitational to atomic accounts of attraction. In fact, at the close of "Attraction" he announces his is a provocative theory and asks if those with the knowledge to prove him wrong will find their way to his column in the *Anglo-African*, implicitly suggesting that they will not.[32] What is gained by substituting the force of electricity for that of gravity? A likely reason that Delany prefers an electrical to a gravitational theory of how systems cohere emerges out of the key claim of Delany's articles on electricity. He proposes that comprehending electricity requires a relational logic: "The terms positive and negative, are simply relative, referring to the comparative condition of each body with the other."[33] He emphasizes that there is no such a thing as an absolutely positive or negative body because any body's properties are defined only in relation to other bodies. His focus on electricity as a relational materialism may well have come from reading Franklin, who helped clarify the confusion that followed on substantializing accounts of electricity by arguing that electricity was not *in* Leyden jars and that its presence, absence, and degree of negativity or positivity emerged through "a network of relations."[34] Delany suggests first that electromagnetic forces are not in bodies but dependent on the relation between bodies and, second, that bodies have very few properties except those that become manifest through their relation. In short, Delany makes relation the origin and cause of centripetal and centrifugal forces, and this pulls the divinity that he also allows into his theory into a relational field. If Delany is interested in offering a relational materialism that focused not on things but on the relay of forces that made things move as they do, this raises another question. Given that gravity is also a relational force and had been recognized as such since Newton, what's the difference between these relationalisms? Eighteenth-century experiments on electricity and the rise of industrial technology made clear that electricity was a relational force that could be manipulated and rerouted by human beings. Gravity could not. Vesting power in human beings and pluralizing what might seem dyadic, Delany's relational materialism suggests that what mattered most wasn't bodies as such but the relations of force by which they held together, moved in some ways and not in others, and sometimes shuttled off on eccentric paths, as Delany claimed comets did.

The relational materialism Delany offers in his electrical theory is apposite to *Blake*, whose first published chapters the Anglo-African's editor, Hamilton, juxtaposed with it. Hamilton casts these chapters as part of a "work" that "shows the combined political and commercial interests that unite the North and the South" and also "gives . . . the formidable understanding among the slaves throughout the United States and Cuba."[35] The book isn't so much a novel but a fictionalized narrative concerning economics, politics, and epistemology, making it a fitting if inadvertent counterpart to *Capital*. *Blake*'s protagonist, who has a plurality of names (one of which is Henry Blake, giving the book its title) is not a single man but becomes a different man in each of the configurations of Atlantic power through which he moves. The chapters published in the *Anglo-African* move among multiple locations, from the plantations of black and white masters (Chapters 7, 17, and 18) to the outlaw territories of the "Wild West" (Chapter 21) to Amerindian nations removed to the frontiers of an expanding U.S. (Chapter 20) to field and forest (Chapter 17) to commercial waterways claimed by steamships and goods (Chapters 19 and 30), and so on. The book tracks how a man whose primary nomination is Blake, though he has many others, takes on different properties and movements when put into contiguity with any given one of these nodes in the Atlantic circuit.

I've suggested that Delany replaces gravitational theory with electrical theory in order to emphasize that relational forces and the systems that they produced might be manipulated. If this is true, we should expect the book not only to track how a body's charges and movements change via its ambit through a relational field but that it should also offer an account of how to manipulate this relational field to produce an alternate circuit of power. This is precisely what Delany's electrical fiction attempts. In each of the nodes of Atlantic power with which he comes into proximity, Blake transmits a secret to a select group of initiates, usually but not always black and always singled out as particularly cagey, so much so as to have an incipient organization even before he arrives to transmit his secret.[36] Delany gives little attention to the content of the secret. Instead, he focuses on the fact that the transmission of the secret to an elect within each node of Atlantic power produces within it a counterpublic that exists in a charged relation to this node of Atlantic power.

For instance, the Mississippi plantation where the book opens presents readers with a key node in the constellation of Atlantic power. In these opening chapters, Blake communicates his secret to two enslaved black men, Andy and Charles, before leaving the plantation to head westward and southward. Andy and Charles remain in place on the plantation to act as a counterforce to the plantation's arrangement of bodies, affects, and knowledge.[37] Communication works differently among those who have power on the Mississippi plantation and the black counterpublic that orbits it. The mostly white plantocracy communicates with each other via letters sent between one plantation and the next as well as to commercial partners in the North.[38] It also communicates through numbers and accounting, evident in discussions and letters concerning Blake's commodity value. Blake is capable of reading, writing, and accounting and does all of these things. For instance, he evidently reads a letter in which his master sells him and then takes action to bypass this economic transaction.[39] This makes clear that Blake deftly navigates the literary cultures that, here, Delany associates with the plantocracy.

However, he, Andy, Charles, and others are also communicating via a non-literary system that allows illiterate blacks on the plantation to transmit private messages among themselves that are not legible to those outside of this countercircuit. When Blake asks Andy and Charles how they knew to meet as well as when and where to meet him, the men respond that they and others on the plantation communicate to each other by the passage of stones: "Ailcey . . . give me the stone, an' I give it to Andy, an' we both sent one apiece back."[40] Surprised by Blake's question, Charles then asks Blake if he didn't receive the stones they'd sent on to him and understand their significance, to which Blake responds that he did receive these stones and this was "the way I knew you intended to meet me."[41] The point of this interlude and the many others like it that occur—in different ways—in each node of Atlantic power is to make clear that the dominant node of power and its means of maintaining the communications networks central to its power (in this case the written word and numbers necessary to buying, selling, and moving black people) has within its orbit another node. This other node is not entirely within the field of the operations of the primary node of power because it builds a charge that puts it in tension with it and that evinces its own circuit of power (in this case communication and knowledge achieved through the exchange of stones). By linking the counterpublics of one node to those of other nodes, Blake's movements attempt to create a black Atlantic countersystem with charge sufficient to overwhelm the circuit of Atlantic power and produce a post-plantation economic, social, and communication system.[42] Delany's serialized novel itself participates in the aim it gives to Blake: The novel elaborates on and gives black leadership to Delany's coadjutor John Brown's plot to trigger a system-wide uprising against the plantocracy. By reading the story, Delany's readers become nodes within the countercircuit of Atlantic power that Delany and Brown attempted to constellate. This last point indicates that literacy cannot be exclusively associated with the plantocracy or Anglo-European communities since it can also be used to create a counter-plantation relation of power (that is, a black counterpublic).

Keeping front and center Delany's relational materialism helps to make sense of *Blake*'s extended treatment of fetishisms. Chapter 24 is a set piece concentrated on the communities created by fetishism (or "conjure," as Delany calls it), which it poses as one node within a black Atlantic circuit of power. Blake comes across and sojourns with the fetishists in the Great Dismal Swamp, which was widely recognized as a location of black resistance in the Atlantic world and one connected to other swamp spaces across the plantation zone. Chapter 24 is one of several chapters where the perspective of a black center of power is prioritized and presented as a force in its own right, not simply as a counterforce to the power of Atlantic capitalism (other examples include the discussion of black creole resistance in New Orleans in Chapter 22 and the black intelligentsia's resistance in Havana in Chapters 60, 61, and 69). That is, while most other chapters, like the opening chapters on the Mississippi plantation, foreground the slavocracy's nodes of power in the Atlantic world, Chapter 24 foregrounds the Dismal Swamp and its fetish-bearing denizens as a key node of black power in the Atlantic world.

Delany's relational materialism requires that no power exist without a counterpower. We can see this relation between a black Atlantic node of power and the counterpower it generates in Delany's extended treatment of one particular fetish in this chapter. One of the men in the Swamp, Gamby, handles what is obviously an Atlantic world fetish:

> He took from a gourd of antiquated appearance which hung against the wall in his hut, many articles of a mysterious character, some resembling bits of woolen yarn, onion skins, oyster shells, finger and toenails, eggshells, and scales which he declared to be from very dangerous serpents, but which closely resembled, and were believed to be those of innocent and harmless fish, with broken iron nails. These he turned over and over again in his hands, closely inspecting them through a fragment of green bottle glass, which he claimed to be a mysterious and precious blue stone got at a peculiar and unknown spot in the Swamp, whither by a special faith he was led—and ever after unable to find the same spot—putting them again into the gourd. . . . This process ended, he whispered, then sighted into the neck, first with one eye, then with the other, then shook, and so alternately whispering, sighting and shaking . . . until finding a forked breast-bone of a small bird, which . . . he called the "charm bone of a treefrog."
>
> "Ah," exclaimed Gamby . . . "got yeh at las'. Take dis, meh son, an' so long as yeh keep it, da can' haum yeh."[43]

This scene does not present fetish use in the swamp in an unequivocally positive light. Instead, it requires that readers engage two oppositional modes of interpretation. First, that at work in Makandal's fetishes, which presents readers with a series of indexical signs: onion skin, oyster shells, human nails, and scales said to be from serpents. Yet this scene doesn't allow readers to straightforwardly interpret these signs from the perspective of the swamp conjurer because readers must also contend with another mode of reading these same signs: that of the narrator, whose passive-voiced, skeptical reading of the fetish artifact plays out the skeptical position of the literate black man reading about this node of black Atlantic power through the vantage of European cultural knowledge gained via literacy. The first reading, that of the swamp conjurer, understands the indexes gathered in the fetish as signs that might be interpreted and reads them through a blue stone. The signs function not only indexically but also iconically, with fish scales designating the snake scales that they resemble. Moreover, the swamp conjurer understands the partial objects gathered in the artifact as emerging from a specific and almost entirely nonhuman matrix to which he was "led" by fate but has never again found but that remains, at least in part, in the stone that indexes it. His scrying of these indexes and icons concludes when he locates what he designates "the charmbone of a treefrog," which he passes on to Blake as protection. From the passive voiced skeptic's perspective, this scene of reading unfolds quite differently. The fish scales that the conjurer links to snake scales are, from the narrator's position, simply fish scales; the charmbone of a treefrog is but the breastbone of a bird; the blue stone the conjurer links to a nearly unknown spot in the swamp is a bit of glass.

These two interpretations of the fetish suggest a tension between an enchanted and what seems to be a disenchanted cosmology and suggests that semiotics functions differently within each. For the swamp fetishists, signs bear the impress of mysterious and only half-comprehended nonhuman significances. For the narrator, such signs are valueless bits and pieces that bear no significance other than that imputed to them by human beings. Indeed, for the narrator and readers like him, the only kinds of significance that signs can have are those that are recognized as having been produced by and for human beings, which is to say an entirely symbolic understanding of language that would attend only to symbolic codes.

Is this scene of fetishism—and the later scenes in *Blake* that reference and replay it—suggesting that Atlantic circuits of power have produced two kinds of signs and with it two kinds of reading, two relations of nature to signs and to value? Put bluntly, does this scene simply present readers with a dialectic? And in staging two apparently antipodal modes of reading, does it simply require that the reader metabolize one or the other of them to develop a more robust understanding of its antipode? This is one interpretation that Delany's writing on fetishisms and the modes of reading they require might warrant. Such an interpretation of fetishism might traditionally have required that signs, hermeneutics, value, and nature as conceived by blacks in the diaspora be incorporated into Anglo-European semiotics, hermeneutics, values and natures. This resolution to the dialectic would allow a kernel of diasporic modes of reading to remain sublated within a rationalized "Western" hermeneutic.[44] It is also possible to flip this interpretation to suggest that Anglo-European signs, hermeneutics, values and natures must be metabolized into diasporic African ones. This resolution would allow a kernel of a rationalized Western style to remain sublated within an enchanted "African" hermeneutic. Neither interpretation, however, captures the dynamic at stake in Delany's relational materialism. The aim of his electrical theory and of his quasi-fiction's application of it isn't to privilege one reading over the other or to metabolize one charge into its antipode. On this point it is important that insofar as *Blake* has an arc it is not plot but the ambit that emerges out of connecting one Atlantic counterpublic to the next. This materializes a hitherto potential, not actual, circuit of Atlantic power that isn't simply orbiting that of a plantation-based Atlantic capitalism but is constituted as its own force field. Both the fetishist's materialized reading and the nature it bears with it, and the rationalist's symbolic reading and its distance from nonhuman nature, would circulate within this countercircuit, which Delany stages so that it includes both of these hermeneutics. This is because, like the positive and negative charges that Delany foregrounds in his electrical theory, the fetishist's and the rationalist's hermeneutics are best thought not as antipodes but as relative charges that become differently positive and negative within any given circuit of power, and whose relations of positivity and negativity shift over time.

In short, Delany's fiction attempts to materialize a constellation of power by playing with, on, and through determining forces. When this constellation of Atlantic power emerges as its own system, it will change and recalibrate the charges of all of the nodes of the system it displaces. This is why we limit our understanding of Delany's project if we take the charges of Atlantic capitalism as absolute and stable charges that congeal into antipodes. The countercircuit of Atlantic power that Delany attempts to build would produce an outside to a plantation-based commodity capitalism. This outside to commodity capitalism Delany attempts to produce emerges mainly in the frontiers of commodity capitalism, which Delany, like Blake, relentlessly traversed in an effort to locate and create counterpublics and then to join them to each other. Once achieved, this frontier would give rise to a fully planetary—indeed, given the literally cosmic scale on which Delany conceives electricity, we might more aptly say cosmological—system that would overwhelm plantation capitalism.

Electric Dialectics and the Problem of History

So far I've argued that Delany's relational materialism doesn't divide out matter from spirit or nature from culture as it attempts to build charged relations between nodes of Atlantic power so as to constellate a post-plantation planetary organization. What, though, is the difference between his relational materialism and Marx's, which I've proposed also rejects the capitalist's division of matter and spirit and argues for relations—namely the class struggle—that would produce a social transformation that exceeds Hegelian sublation? The colonial frame, which gives rise to the phenomenon of fetishism, orients Delany's relational materialism differently than Marx's: first, because this frame adumbrates a different planetary order than that imagined from the perspective of Europe, and second, because this frame requires a different relation of the past to the present, and with it a different account of spirits and spiritualism.

On this point, consider one of Marx's most famous ghost stories, *The 18th Brumaire of Louis Bonaparte* (1852), first published in the New York German-language monthly *Die Revolution*. This pamphlet diagnoses the problem with existing revolutionary movements, which have culminated in the decidedly counterrevolutionary ascendance of the bureaucratic, bourgeois republic of Louis Bonaparte, and it also articulates the orientation necessary to the socialist revolution to come. Marx criticizes previous revolutionaries who have hidden from themselves the radicalism of their own ambitions by cloaking their actions in the drag of earlier historical moments; the socialist revolution to come must strip itself of such historical drag and "draw its poetry . . . only from the future."[45] Here Marx would seem to offer a simple opposition between revolutionaries who blunt their power by suffusing their movements with the ghosts of the past and socialist revolutionaries whose power will come from looking only to the future. Yet Marx doesn't quite oppose the past to the present or superstition to the forward-looking scientism of socialist revolutionaries. Against the banality of Louis Bonaparte whom Marx derisively claims could never be cast as a sorcerer (*Hexenmeister*), spiritualism and ghosts retain and carry into the present the germ of the messianic promise that the socialist must awaken.[46] What's more, in Marx's analysis, the Americas, particularly the United States, are in a quite peculiar relation to this process of awakening a germ of the past to catalyze the socialist revolution to come: The Americans' frenzied material production has left them "neither time nor opportunity for abolishing the old spirit world (*Geisterwelt*)."[47] On the one hand, the Americas evince a total presentism that allows for no history; on the other hand, in having no history, they have not abolished the specters of the past, and for this the Americas are nothing but a spirit world.

Here, again, Marx's comments on the Americas are made passingly and for Marx the socialist revolution to come will move from the old world to the new. Yet Delany's relational materialism allows us to consider how to put together the colonial condition only at the frontiers of Marx's analyses with the Americas' peculiar historicism, in so doing reversing Marx's claim that the new world will follow on the old. Delany's relational materialism turns his readers to the pasts that circulate in the present, training their attention on the fetishisms that emerged as a consequence of colonialism and that persisted in a later moment whose social and economic relations were still structured by the plantation form in particular. These fetishisms circulating through the nineteenth century cannot be taken as a historical past being reanimated in the present because the colonial crises that occasioned them are not recorded as history or, insofar as they do pass into history, remain only as details or ephemera. This, too, because these crises and practices were not past but continue at the edges of the plantation form in the nineteenth-century present. The relational materialism I've culled from Delany allows us to stay within and rethink the degraded and cast off as central to locating alternatives to capitalistic modes of circulation and valuation. It also indicates the importance of the degraded and cast off to a non-monumental historicism attentive to what had no place in history proper.

Now, too, we might need encouragement as we sleepwalk through a moment that seems to mark the triumph of a neoliberal capitalism that has eaten all of nature. It's time, again, to be cheered by the strange movements on the edges of empire and the histories and materialisms that flash forth from them. Where does the Atlantic fetish still brush against us? How can we take up Delany's work of joining one strange materialism to the next so that we might not be isolated islands of resistance but an archipelago with ambit and charge enough to make the present not only livable but the future of the past that never was?

Notes

1. Karl Marx, *Capital*, vol. 1., trans. Ben Fowkes (New York: New World Paperbacks, 1967), 163–64.
2. The reference to China is to the Taiping Rebellion (1850–1864).
3. Marx and Engels were not precisely opposed to phenomena like mesmerism. A friend of Marx's warned him against his interest in the "charlatanry of mesmerism, for which you have always had a weakness." See Hal Draper, *Karl Marx's Theory of Revolution*, vol. 4, *Critique of Other Socialisms* (New York: Monthly Review Press, 1990), 255. Engels and a Manchester acquaintance attempted and succeeded in replicating mesmeric phenomena, proving that mesmeric trance was real even if the conclusions bourgeois scientists drew from this (the existence of God) were wrong. See Frederick Engels, *Dialectics of Nature* (New York: International Publishers, 1940). If the most obvious reason for Marx's interest in spiritualisms is that they are materialisms, another might be that mesmerism and spiritualism were advanced by radicals who broke with bourgeois familial and state forms.
4. See Howard Kerr, *Mediums, and Spirit Rappers, and Roaring Radicals: Spiritualism in American Literature, 1850–1900* (Urbana: University of Illinois Press, 1973), 3–20.
5. Harriet Beecher Stowe, *Dred* (New York: Penguin, 2000), 274. On mesmerism in *Dred* see Christina Zwarg, "Who's Afraid of Virginia's Nat Turner? Mesmerism, Stowe, and the Terror of Things," *American Literature* 87, no. 1 (2015): 23–50. On Stowe's spiritualism, see Alex Owen, *The Darkened Room* (Chicago: University of Chicago Press, 1989), 19–20.
6. See especially Douglass's letters to Rochester-based abolitionist and spiritualist Amy Post in *The Frederick Douglass Papers: 1842–1852*, ed. John R. McKivigan. (New Haven: Yale University Press, 2009). See also the April 5, 1850 column on spiritualism in the *North Star*.
7. Jason W. Moore, *Capitalism and the Web of Life* (New York: Verso, 2015).
8. Christopher Columbus, "Letter to the King and Queen of Castile," in *The Heath Anthology of American Literature*, concise ed., ed. Paul Lauter (Boston: Wadsworth Publishing, 2007), 13–14.
9. William Bosman, *A New and Accurate Description of the Coast of Guinea: Divided into the Gold, the Slave, and the Ivory Coasts* (London: Frank Cass and Company, 1967), 12–13.
10. William Pietz, "The Problem of the Fetish I," *RES: Anthropology and Aesthetics* 9 (1985): 5–17.
11. Sébastien-Jacques Courtin, "Mémoire sommaire sur les prétendues pratiques magiques et empoisonnements" (1758), Archives Nationales Colonies (Paris) F3 1758.

12. Ibid.

13. Karol K. Weaver, *Medical Revolutionaries: The Enslaved Healers of Eighteenth-Century Saint Domingue* (Urbana: University of Illinois Press, 2006), 66.

14. Reliquary in Taino and Arawak burial is discussed in Fray Ramon Pané, *An Account of the Antiquities of the Indians*, ed. José Juan Arrom and trans. Susan C. Griswold (Durham: Duke University Press, 1999); Père Nicolson, *Essai sur l'histoire naturelle de l'isle de Saint-Domingue* (Paris: Gobreau, 1776); and Kathleen Monteith and Glen Richards, *Jamaica in Slavery and Freedom: History, Heritage and Culture* (Kingston: University of the West Indies Press, 2000).

15. Suzanne Preston Blier, *African Vodun: Art, Psychology, and Power* (Chicago: University of Chicago Press, 1995).

16. Charles Peirce, "What Is a Sign?," in *The Essential Peirce*, vol. 2, *Selected Philosophical Writings (1893–1913)* (Bloomington: Indiana University Press, 1998), 4–10.

17. Amerindian ontologics may also have presumed the imbrication of nature in culture. See Philippe Descola, *Beyond Nature and Culture*, trans. Janet Lloyd (Chicago: Chicago University Press, 2013).

18. Weaver, *Medical Revolutionaries*, 100–12; James McClellan, *Colonialism and Science: Saint Domingue in the Old Regime* (Chicago: University of Chicago Press, 1992), 178.

19. F. A. Mesmer, *Memoir of F. A. Mesmer, Doctor of Medicine, On His Discoveries*, trans. Jerome Eden (Mt. Vernon: Eden Press, 1957), 10, 11.

20. McClellan, *Colonialism and Science*, 178.

21. Weaver, *Medical Revolutionaries*, 108.

22. Cited in Weaver, *Medical Revolutionaries*, 112.

23. Jared Hickman, *Black Prometheus: Race and Radicalism in the Age of Atlantic Slavery* (New York: Oxford University Press, 2016).

24. For Delany's biography, see Dorothy Sterling, *The Making of an Afro-American: Martin Robison Delany, 1812–1885* (New York: Doubleday, 1971).

25. Delany departed from that paper and Douglass's circle in part because of his more radical and less cosmopolitan positions on subjects ranging from racial blackness to constitutionalism to colonization.

26. Britt Rusert was the first to note the conjunction between Delany's fiction and what she calls his "science" and what I call his materialism; my own reading is indebted to her wonderfully productive analysis. See Rusert, "Delany's Comet: Fugitive Science and the Speculative Imaginary of Emancipation," *American Quarterly* 65, no. 4 (2013): 799–829.

27. See Robert Levine *Martin R. Delany: A Documentary Reader* (Chapel Hill: University of North Carolina Press, 2003), 25; and Sterling, *Making of an Afro-American*.

28. Ann Braude, *Radical Spirits: Spiritualism and Women's Rights in Nineteenth-Century America* (Boston: Beacon Press, 1989), 23–25, 27–29. For more on the relationship between race and electricity, see James Delbourgo, *A Most Amazing Scene of Wonders: Electricity and Enlightenment in Early America* (Cambridge: Harvard University Press, 2006).

29. Braude, *Radical Spirits*, 4–5. The appropriations bill in favor of funding Morse's telegraph project only very narrowly passed.

30. Martin Delany, "Attraction," in *The American Negro, His History and Literature*, vol. 1, *The Anglo-African Magazine* (New York: Arno Press, 1968), 17–20.

31. George Vashon, "Successive Advances of Astronomy," in *The American Negro, His History and Literature*, vol. 1, *The Anglo-African Magazine* (New York: Arno Press, 1968), 208.

32. Delany, "Attraction," 20.

33. Ibid., 18.

34. David Morse, *Perspectives on Romanticism: A Transformational Analysis* (London: Macmillan, 1981), 99–100.

35. Delany, "Attraction," 20.

36. Martin Delany, *Blake: Or, The Huts of America* (Boston: Beacon, 1970), 91 (for an example of the text's silence on the substance of Blake's secret); 89 (for an example of a black community with an incipient organization before Blake's arrival).

37. Delany, *Blake*, 36–44.

38. Ibid., 24, 27.

39. Ibid., 29, where Blake knows the sum for which he was sold to Harris.

40. Ibid., 36.

41. Ibid.

42. Delany collaborated with Brown in Canada, and his absence from Brown's attack at Harper's Ferry was because he was in Africa when it occurred. See Sterling, *Making of an Afro-American*, 167–75.

43. Delany, *Blake*, 112–13.

44. By my reading, this is the interpretation that Rusert gives to this scene in "Delany's Comet," 818.

45. Karl Marx, *The Eighteenth Brumaire of Louis Bonaparte* (New York: International Publishers, 1994), 18.

46. Ibid., 20.

47. Ibid., 25.

CHAPTER 11

Satire's Ecology

Teresa Shewry

> corrugated tracks stepping up grass
> to the farm, the hills, the Maoritanga, the town,
> all bundled together with number eight wire,
> and dumped on the tray of a driverless ute,
> revved up on the last of the petrol,
> and spluttery, like water tanks drained
> in late summer, leaving the taste of grinding
> peppermill dust and dry forest floor.
>
> —DAVID EGGLETON, "Driverless Ute"

Satirical humor, as called up in David Eggleton's poem "Driverless Ute" (2010), seems to respond to, and then be exhausted by, ecological depletion. In this work, a "ute," or utility vehicle, emerges in an oceanic place marked by constraint and endings: the last petrol, exhausted water tanks, waves "strait-laced in pews."[1] An enigmatic figure, this ute is driverless and carries the very terrain through which it travels: A farm, hills, Māoritanga (Māori culture), and a town are "all bundled together with number eight wire" and "dumped on the tray."[2] The ute moves in and bears a broader material world; in so doing it suggests how forms of life involving petroleum draw on but also shape the changing Earth. It exaggerates a New Zealand discourse, sometimes known as kiwi ingenuity, that valorizes an easy-going attitude and cunning inventiveness in the face of difficulty.[3] The ute takes this discourse to extremes in suggesting that even as things are falling apart everything can be thrown together and "we'll" continue onwards, without collective discussion about direction, what and who is taken, and the viability of understanding of life as a kind of driving, and with continuing attachment to fossil fuels. Eggleton uses exaggeration to spotlight this discourse as amusing, inadequate, and, indeed, as frightening in the face of climate change. At the end of the poem, the ute struggles as its petrol runs out, "like water tanks drained," a somber image of depletion that also suggests a different, perhaps incalculable continuation.[4] We move from the satirical figure of this sad vehicle, rendered in the past tense until a last vision of it as "spluttery," to a final, somber present-continuous tense image of dust and desiccation.[5] Satirical humor emerges here with its source and target in ecological loss before seeming to run out or deplete, too, as if Eggleton cannot imagine his own mode, satire, having a future.

To laugh would seem to be impossible, or to bring its own violence by making a mockery of urgency and devastation, in relation to climate change and drought, among the concerns of "Driverless Ute" and of other poems in Eggleton's *Time of the Icebergs* (2010). An Aotearoa, New Zealand, poet of Rotuman, Tongan, and European heritage, Eggleton has long written cutting poetic satire about economic and ecological life in the Pacific and beyond. But in "Driverless Ute," satire is itself figured as spluttering, an ending that is not quite over. Eggleton suggests that there may have been a time when satire was more alive or more possible, orienting us into the past. Satire's haunting past might, in turn, enliven our understanding of these humorous formal techniques that falter, perhaps because read as violent or inadequate, in a present moment so heavily circumscribed by crisis. I begin this chapter, then, with Samuel Butler's *Erewhon* (1872), an archive of another journey from ocean to mountains, also loosely evoking New Zealand, from a moment when literary satire was being connected to the furious ecological violence of the British Empire. First imagined in New Zealand in the 1860s, published in London in 1872, and revised in 1901, *Erewhon* deploys satirical forms—including irony, sarcasm, and caricature—to blast settler-capitalist approaches to humans and to other life forms and elements as both violent and avoidable. In suggesting satire's potential to undertake sharp, if unstable, structural critique and to make space for utopian insistence that this world could be different, *Erewhon* eventually turns me back to "Driverless Ute," to consider satire in relation to climate change and particularly to explore the apparent tensions between Eggleton's satirical critique of current economic and social life, with its evocation of openness to maneuver, and his final, bleak images of endings and of continued ecological depletion.

Precarious Humor

An Indigenous man "drowns" the English narrator of *Erewhon* in a river from which the body could never be retrieved. So we are told, toward the end of this work, of events that unfold after the narrator leaves a settler colony and follows a river into its mountain headwaters in search for land. He takes Chowbok (at times rather named Kahabuka and Rev. William Habakkuk) as guide and to lead his packhorse. Chowbok flees the narrator after they spend weeks struggling up a river gorge, while the narrator continues through difficult terrain and is surprised to find a utopian society named Erewhon. The Erewhonians trap him, but he eventually escapes to London, at which point we learn that Chowbok had returned to the settler colony and assumed that the narrator would never make it back. He "made up a story about my having fallen into a whirlpool of seething waters while coming down the gorge homeward. Search was made for my body, but the rascal had chosen to drown me in a place where there would be no chance of its ever being recovered."[6] Butler uses humor here to navigate complex, disturbing relations involving an English settler-narrator, an Indigenous man, and water. The narrator's fictive drowning broadly reflects what many critics observe as *Erewhon*'s satirical thematic and formal techniques.[7] Robert C. Elliott notes that the concept of satire has come to be detached from a single genre and now signals a work that contains "a sharp kind of irony or ridicule or even denunciation" and that has a "derisive or sarcastic" tone.[8] As Angelique Haugerud writes, in its expansive modern sense, satire attacks through "ridicule, parody, or caricature."[9]

Butler links satirical humor to the narrator's effort to discover land already within the "traditions" of Chowbok's people, to a deeply troubled settler connection with rivers, and to Chowbok's possible affective and imaginative engagement in empire, partly defusing these concerns as an amusement.[10] But the history of satire suggests that we can read it for bleaker, more turbulent potentials. As many critics note, *Erewhon* participates in long-standing, arguably inextricable, connections between the utopian literary form and satire.[11] Of the "idiotic names" in Thomas More's *Utopia*—a river named Anydrus, or "no water," for example—Fredric Jameson notes that utopia's satirical techniques can allow for it to be read as "a *jeu d'esprit*," drifting free of both serious critique and imagined alternatives.[12] Drawing on Robert C. Elliott, however, Jameson argues that satire might be understood not as rejecting utopian imaginings of expansive potentials, such as the abolition of private property, but rather as "the passionate and prophetic onslaught on current conditions and on the wickedness and stupidity of human beings in the fallen world of the here and now."[13] Indeed, Elliott argues that ancient Greek, Irish, and Arabic archives suggest that people may have understood satire's onslaught to exert a "malefic power" that could be deadly for its targets.[14] This adverse force could affect not only people but broader environments: "Some of the great Celtic satirists were able to blight the land itself—a curious reversal here of the function of satire in the rituals of Greece, where it promoted fertility."[15] Elliott suggests that satire's critical approach can nourish utopian engagement with the contingencies and possibilities of the present world: "The very notion of utopia necessarily entails a negative appraisal of present conditions. Satire and utopia are not really separable, the one a critique of the real world in the name of something better, the other a hopeful construct of a world that might be."[16] In destabilizing common assumptions and understandings, satire may also help make thinkable less-recognized potentials of utopian literary form, including the illumination of scalar shifts within the systems where human and nonhuman beings interweave, as against hegemonic imaginaries of autonomous individuality, as described by Benjamin Morgan in this collection.[17] Many understandings of the violence and promise of satire have been lost as the term underwent "enormous inflation of meaning" during its travels from late antiquity.[18] The broad scale of satire's histories suggests, nevertheless, that its humor may have complex, turbulent implications.

We might start, following Jameson, by considering what could be read as "passionate" in Butler's story: the potential deadliness of the "seething" waters, and Chowbok's imaginary of a settler-narrator who drowns and disappears forever.[19] *Erewhon* reflects the importance of rivers for navigation and sustenance in the British Empire. Butler, who was born in England in 1835, drew *Erewhon* from articles he published between 1860–1864 while living in Canterbury Settlement in New Zealand. He claims to have modeled the narrator's and Chowbok's journey on the upper reaches of the Rangitata River.[20] This river cuts through immense floodplains that took their present form over thousands of years as water and sediment washed off glaciers and through flooding and sediment from braided rivers that emerge in the Southern Alps and make their way to the ocean. The iwi (tribe) Ngāi Tahu maintained settlements and migrated to gather food in the catchment and on the floodplain. The river was interwoven in their routes from East to West Coast.[21]

Almost as soon as the narrator and Chowbok set off from the colony in *Erewhon*, the river's flooding dynamics emerge. "We knew that it was liable to very sudden and heavy freshets," the narrator tells us, "but even if we had not known it, we could have seen it by the snags of trees, which must have been carried long distances, and by the mass of vegetable and mineral *débris* that was banked against their lower side."[22] Such references provide a memory of ecological relations that have been all but decimated. The

Figure 5. Satellite map of the Rangitata River amid grids of farmland across the floodplain. (Image courtesy of Google, DigitalGlobe.)

snagged trees that our narrator describes suggest the presence of riparian and catchment forest. If Butler was, indeed, thinking of the Rangitata when he wrote this section of *Erewhon*, the reference is likely drawn from the podocarp forest—kahikatea and matai trees—that once dominated the river's swampy floodplain.

In satellite imagery, the lower Rangitata appears as a series of shaggy, intersecting braids that fail to conform to the vast flat field of green geometric grids across the floodplain. The river seems untouched within this startling contrast, despite the evident upheaval of its surrounds (see Figure 5).

Yet rivers and their floodplain ecosystems have been subject to some of the most extreme ecological violence of the British Empire in New Zealand.[23] From the 1840s, settlers cut and burned tussock and bush, drained swamps, and privatized land on the Canterbury Plains. They integrated the plains into an extractive economy based on crops, sheep, and cattle, expanding their cultivated land from 292,950 acres in 1871 to more than 2,191,185 acres by 1895.[24] They also set out to separate the rivers from the floodplains by engineering stop banks, groynes, and channels, intensifying the siege on life forms such as kahikatea, a tree that relies on swampy terrain for long-term survival. "You can find it in the hills, but it only prospers in the swamps. It would vanish without them," writes ecologist Geoff Park.[25] Today it is thought that less than half a percent of the Canterbury Plains supports native plant communities.[26]

The end of forests is a concern that shapes New Zealand poetry around the turn of the twentieth century. Fragmentary references to destroyed or altered freshwater ecosystems also mark the literature of the time: a road that was once a stream, willows growing on riverbanks.[27] But writers struggle, or refuse, to comprehend the destruction of forests as violence that spreads to water. In "A Bush Section" (1908), for example, Blanche Baughan mobilizes a river as a site of exciting movement in a landscape heavily characterized by endings and stillness. An orphan child named Thor lives on a settler farm surrounded by remnants of felled and burned trees, a landscape "ruin'd, forlorn, and blank."[28] Baughan suddenly breaks this forlorn narrative with questions: "What glimmers? What silver/Streaks the grey dusk?"[29] The river, a survivor, always on the move, affirms change and suggests that the world still has momentum and a future. Figured as Thor's "playmate, his comrade," the river embodies a companionable relationship between environment and settlers, something that Baughan cannot extract from the hacked, charred remains of trees.[30] Drawing on the river, a train, stars, Thor, and finally "the resolute Settler" as evidencing the inevitability and rectitude of change, Baughan tries to sweep us up in a glorious "dawning" at the hands of settler colonialism's evident ecological devastations.[31]

The amphibious dynamics of river and floodplain were not lost on all writers, however. William Pember Reeves, born in Canterbury in New Zealand, writing in 1889, provides a rare elegiac account of a swamp. He describes its *"sweet and clear"* water, the flax, toetoe, rushes, and other vegetation that inhabit it, and the violence of draining and agriculture: *"the peat sank, cracked and dried, the surface was systematically burnt and became stretches of black, hideous ashes and mud, poached up by the hoofs of cattle."*[32] Reeves's is an unusual voice in a culture that more often took swamps as horrific, wasteful, and sad entities only good to be drained. He provides an archive of a lost ecosystem, in this part of the world where swamps are now reduced by around 85 percent of their extent at European settlement. But Reeves is only willing to draw us so far beneath the surface of these murky waters. The swamp is "to be admired in its vanishing," as ecologist Geoff Park puts it.[33]

Precise distinctions between land, water, and life, and upheavals to these complex relationships, continue to percolate in Reeves's later work,

"The Passing of the Forest" (1898). In this influential poem, he describes the destruction of the forest "world" and "nation"—provocative terms in the context of settler colonialism—attributing vibrancy and autonomy to the forest while also expressing a modern redemptive story that aligns violence with progress.[34] As Philip Steer notes, Reeves "uniquely invokes the complexity and diversity of the Indigenous ecosystem."[35] The narrator recalls that he used to hear a river in the valley far below while riding through the forest, its "voice through many a windless summer day/Haunted the silent woods, now passed away."[36] He remembers a waterfall that ran down the hillside: "White, living water, cooling with its spray/Dense plumes of fragile fern, now scorched away."[37] Reeves describes the interconnection of water and fern, but remains ambiguous on whether the waterfall and the river were destroyed or damaged along with the vegetation. Water is elusive in relation to the colonial aesthetic of environmental loss. Yet the destruction of forest, and the agriculture that follows, profoundly alters rivers, swamps, ground water, and coastal ecosystems, including through increased siltation, as soil slides off clear-cut land, and heightened flooding, as without trees the land can no longer soak up as much water.

In Reeves, settlers' emergent literary engagement with environmental destruction overlapped the early interventions of the colonial government. In the 1890s, the government was putting in place infrastructure to respond to the loss of forest, with a focus on preserving scenery.[38] The Land Act (1892), for example, allowed the reservation of land to protect scenery of national significance. Reeves became involved in these politics as New Zealand's first Minister of Labour from 1892 to 1896. Concerned about the concentration of land in the hands of a few, he and Edward Tregear, another writer and a public servant, experimented in establishing a cooperative farm for destitute peoples. The government acquired the land for the farm from Warena Hunia, a Muaūpoko chief, pushing him to give up the land so he could clear debt.[39] Geoff Park writes that Reeves and Tregear felt conflicted, not about the violence of the land transaction but that their "State Farm" experiment would be "part of the relentless campaign against nature" because "giving the 'landless' what wealth had given the landlord" would involve allowing "more men to 'strip the woods away.'"[40] Their plan to locate the settler poor in the area also came into tension with naturalist Walter Buller, owner of a decadent neighboring country estate. Reeves and Tregear navigated these issues by protecting a swampy forest alongside the State Farm. This fragment of native ecosystem is known as Papaitonga Scenic Reserve.

Placing this bureaucratic context alongside "The Passing of the Forest" reminds me of how the poem's form of environmental destruction is one imbued with authority, from Reeves's somber account of tragedy at an immense scale to his evident desire to make a final definitive statement about the meaning of this destruction as "progress."[41] One could tell the story of the forest differently: Reeves, for example, simply makes Māori vanish along with the trees; moreover, he states that the forest was taken for progress instead of, say, the capitalist economy. But it is difficult to extract much sense of interpretative potential and tension from the poem. The integrity of Butler's narrative in *Erewhon*, in contrast, seems always to be falling apart or on the verge of doing so, as the satire undermines the authority of the narrator and our readings. There are continuities in Butler's and Reeves's environmental aesthetic, including in Butler's appreciative descriptions of mountain and riverine scenery in the early chapters of *Erewhon*. But humor is entirely absent from the "The Passing of the Forest." Reeves's sober form of environmental loss, in turn, might give pause regarding *Erewhon* and the potential implications of a literary form that, at some level, seems crafted to encourage laughter in relation to an environmental life heavily characterized by crisis and destruction.

Irony and Ecology

Rivers, along with many elements and beings, enter *Erewhon* as sites of wild economic aspiration. Almost as soon as *Erewhon* begins, Butler sets about rebuking an extreme European orientation toward accumulating money. Our narrator, in a rare perceptive mode, tells us that a woolshed in the settler colony is "built somewhat on the same plan as a cathedral, with aisles on either side full of pens for the sheep, a great nave.... It always refreshed me with a semblance of antiquity (precious in a new country), though I very well knew that the oldest wool-shed in the settlement was not more than seven years old."[42] He mocks settlers' lack of history in the colony, but the deeper target of this taunting sarcasm is the extremity of their worship of capitalist enterprise and their failure to see life in any other way, as well as their dressing up of economic motives in spiritual terms. The narrator provides a direct, critical voice here, but he often participates unthinkingly in such economic devotion and is caught up in "uncritically recording or even embracing the folly which it is the satirist's business to undermine."[43] In the woolshed scene, indeed, he is trying to persuade Chowbok to take him up the river so that he can grab land, or gold, diamonds, copper, silver, slate, and granite if land is not available.

Such moments of satirical humor explode the seeming naturalness and inevitability of their target by throwing it into view in playful and unexpected terms. In so doing, they bear the potential to "perform dissent in a way that surprises and charms," to draw from Angelique Haugerud's discussion of satirical activism in the United States.[44] To suddenly dramatize an unacknowledged settler devotion to the capitalist economy as *funny* in its extremity is to open space to express disquiet. Satire's humor can disrupt ordinariness, solemnity, and authority, allowing for critique. The woolshed was a vital infrastructure of settler colonialism as a hinge between the industrial farming of sheep and a transnational trade in wool. But in some forums the woolshed was coming to be comprehended as reflecting national identity, not as a symbol of participation in a global economy, as Butler implies in mockingly noting that these infrastructures are a settler tradition.[45] As *Erewhon* proceeds, Butler suggests, over and over, that the narrator's consuming devotion to such an economy involves a deeply callous and precarious approach to relationships with others, from his arrogant engagements with Chowbok, to his figuring of ecosystems that cannot be farmed as worthless, to his plan to forcibly transport the Erewhonians to labor on Queensland plantations.

Satire responds to violence, then, but it also gives form to a different imaginary of the life and potential in which human and nonhuman beings are interwoven. Butler heavily wields irony, in particular, to illuminate dynamic and disturbing relationships between the narrator, other people, and other beings. An ironic narrative, writes Haugerud, turns us toward "a state of affairs contrary to what might be expected."[46] In the story of Chowbok drowning the narrator, Butler opens up multiple faultlines between what we may perceive to be unfolding and what characters understand or say is unfolding. The narrator, of course, says something different from what he means, as he recounts his drowning at the hands of Chowbok. He appears to intend to use this story to position Chowbok as a comic "rascal."[47] Yet, he also opens up interpretative possibilities that exceed his grasp. The story is underwritten by layers of what Haugerud calls "cosmic irony."[48] Haugerud quotes from Claire Colebrook's description of such irony as a theorization that "refers to the limits of human meaning; we do not see the effects of what we do, the outcomes of our actions, or the forces that exceed our choices."[49] This approach to understanding and intent is evident in the settlers' futile search for a drowned man. It touches on Chowbok in his speculation that the narrator will not come back from the mountains.

But the main target of Butler's ironic humor is the narrator, who is imaginatively drowned in the river that is supposed to be his pathway to realizing expansive wealth and fame, by the man he plans to convert so as to reach spiritual heights. At the beginning of *Erewhon*, the narrator tells us that he wants to use rivers to navigate into unexplored terrain, grab land, and "secure me a position such as has not been attained by more than some fifteen or sixteen persons, since the creation of the universe."[50] Near the end of the novel, he again reports his plans to use a waterway to advance these aspirations, stating he will travel by gunboat into Erewhon and transport the Erewhonians to sugar cane plantations in Australia. But the narrator's relationships with other humans, in particular Chowbok, and with other life forms and things, including water, complicate and even hold the potential to destroy these aspirations. Butler's ironic approach situates agency as interconnected, precarious, and as having implications not fully understood. He suggests that there is no straightforward pathway from aspiration to a given future, as plans will be disrupted by other beings, human and nonhuman. He also emphasizes that actions may have implications beyond what we grasp. The narrator might be setting in place the conditions for his own drowning, rather than for achieving wild wealth and fame. As *Erewhon* unfolds, we must recast our understanding of what the narrator has been doing: Instead of moving into appropriable land, he has been walking toward Erewhon.

Positioning others as lively, unpredictable agents can have unstable implications, serving colonial narrative designs in which a European pioneer is severely tested but triumphs. In William Pember Reeves's poem, "A Colonist in His Garden" (1898), for example, a settler must "Fight Nature for a home," facing storms, drought, and flooding to establish a garden where an "English rose" (that is, his daughter) can walk.[51] Butler's irony also emphasizes the complications of living with other elements, but it offers a very different understanding of the realities of such relationships and orientation into their futures. It undermines the possibility of imaginatively resolving the precariousness bound up in living with others, a closure that Reeves's narrator seemingly achieves in finally establishing a garden. It also replaces Reeves's steady positioning of storms, droughts, and floods as antagonists with emphasis on limitations in understanding. In *Erewhon*, water does not so much resist the settler as form part of a complex material world that interacts with the narrator's idiocy, arrogance, and violence: Sometimes the river is a source of food and a pathway; sometimes it is "horribly angry."[52] These relationships never resolve into progress or any other fiction. At the end, we are faced with the narrator's plan to return to Erewhon, from the vantage point of having repeatedly seen his plans going awry before our tired eyes. As Haugerud suggests, satire is distinct from comedy in refusing resolution: "Satire . . . assumes that the pleasing resolutions of comedic narratives are inadequate."[53] Every time our narrator is surprised, he renews his effort to capitalize on whatever he can, with relative detachment from particular places and engagements, aside from accumulating wealth. When he does not find appropriate land, for example, he formulates a new plan to transport the Erewhonians. This flexible capitalist character is a reminder that satire's critical work, its emphasis on the need and potential to change economic and political structures, is vital alongside the imaginary it offers of a potentially disturbing, complex environment as an alternative to the narrator's claims that the land is worthless.

Butler also uses satire to reveal instabilities in relationships between people, and in particular to expose and loosen the hold of the narrator's assumption of supremacy in relation to Chowbok. In telling the story of the narrator's drowning, Chowbok apparently imagines the narrator disappearing permanently. He never directly expresses his imaginative and emotional relationships with the narrator and with empire, but Butler makes oblique references to the complexity of such relationships, taking us beyond the narrator's narratives of him as a stupid, barbaric, comic figure. For example, by fictively drowning the narrator, Chowbok circumvents having to answer questions about where the narrator might have gone. He does not want the narrator to cross further into the mountains or to be drawn there himself; he seems to know that something horrific lies ahead. Early in the novel, the narrator's questions about the mountains beyond the settlement make Chowbok uneasy: "I could see that of this too there existed traditions in his tribe; but no efforts or coaxing could get a word from him about them."[54] Chowbok later seeks to undermine the narrator's effort to travel up a stream onto the mountain range, stating that he has already checked to see whether the stream could be navigated and that this would be "impossible."[55] When the narrator heads up the stream regardless, finding it to be navigable, Chowbok refuses to follow, returns to the settlement, and tells the settlers that the narrator has drowned. The narrator, meanwhile, feels that "Chowbok had designedly attempted to keep me from going up this valley."[56] An explanation for Chowbok's obstruction of the journey onto the mountains is eventually offered when the Erewhonians tell the narrator that they once used Chowbok's people as sacrifices and that if they were to enter Erewhon in the present day they would be confined and deliberately bored to death.[57] Butler suggests Chowbok's canny maneuvering in the imaginative and material life of empire and perhaps also hints at grievances. He disrupts the authoritative logic of the narrator's stories of Chowbok, positioning them among the narrator's efforts—extending from ideology to material force—to maintain control of a contingent social hierarchy.

But in the character of Chowbok, the precariousness of Butler's satirical humor also becomes particularly evident. As I have noted, the narrator casts Chowbok as a "rascal" for telling a lie about his drowning, minimizing Chowbok's act as laughable mischief.[58] In figuring Chowbok as a troublemaker, and a non-threatening one at that, he deflects attention from the deeper issue of the legitimacy of the colonial designs that are being troubled. Our narrator relentlessly positions Chowbok as a figure to be laughed at, in ways layered into *Erewhon*'s critiques of European extremity, shallowness, and incomprehension. His insistence that we laugh at Chowbok as a rascal, for example, draws attention to this scene and may orient the reader into its deeper critical meanings. *Erewhon* seems to allow for both "laughing *at* and laughing *with* those in power."[59] No one escapes its satire. Yet humor plays out in conditions of socioeconomic unevenness. When targeting a settler here, humor may be readable as intriguing dissent, but turned against Indigenous peoples, it may take part in an ordinary, chronic system of undermining. Although a tragic narrative of Māori as a disappearing people appears in European New Zealand archives of this time, we also find imperatives to laugh. For example, a 1906 comic in the *Auckland Weekly News* mocks Māori claims on activities that, for European cultures, bear deep cultural prestige, such as reading (see Figure 6).

Something of this violent history of humor is evoked in Te Whānau-ā-Apanui, Ngāti Porou, and Taranaki writer Apirana Taylor's "Sad Joke on a Marae" (1979). In this poem, the narrator approaches a marae, describing himself as "Tu the freezing worker/Ngati D. B. is my tribe."[60] He expresses his relationships with a slaughterhouse, alcohol ("D. B." is a common New Zealand beer), the pub, violence, and jail, as well as what Alice Te Punga Somerville calls his "reciprocal relationship" with the beings of this place.[61] Taylor writes of these beings, "They understood/the tekoteko and the ghosts."[62] The phrase "sad joke" gives expression to the feeling of being a joke, and one that is not funny at that, in conditions that include economic impoverishment and alienation from Māori language. Yet the "sad joke" may also evoke European caricatures of Māori men—such as Butler's figuring of Chowbok as desperate to acquire alcohol from the narrator—as joking about deeply unjust economic and social conditions in a way that targets Māori as comic failures while deflecting attention from how such conditions have been shaped by colonial history.

Satire's Ecology   235

Figure 6. Racist humor. A 1906 *Auckland Weekly News* cartoon by Trevor Lloyd mocks Māori claims on activities that for European cultures bear cultural prestige, such as reading. (Image courtesy of Sir George Grey Special Collections, Auckland Libraries, item AWNS-19060125-10-2.)

At the end of *Erewhon*, our narrator expresses that the story he has been telling us—*Erewhon*, presumably—is actually an advertisement for his scheme to forcibly transport Erewhonians to Australian plantations. Butler turns the force of satire onto his own writing, positioning *Erewhon* as a call for participation in extreme violence:

> I can see no hitch nor difficulty about the matter, and trust that this book will sufficiently advertise the scheme to insure the subscription of the necessary capital; as soon as this is forthcoming I will guarantee that I convert the Erewhonians not only into good Christians but into a source of considerable profit to the shareholders.[63]

We might remember that Chowbok drowns not simply a settler but a narrator. Here we catch a glimpse of the silences and inadequacies meant by reading literature as an archive. Satire may destabilize sedimented ideas, undertake fierce critique, and express different imaginative possibilities, but in its persistently dissatisfied relationship with understandings and engagements, it can turn us critically onto inadequacies in its own commitments.

Satire's Futures

Butler's sense of a present world not adequately comprehended takes new meaning when juxtaposed with Eggleton's "Driverless Ute," written more than one hundred years after *Erewhon*. What Butler could not know was that industrial agriculture, one of the targets of his satire, profoundly shapes the Earth through greenhouse gas emissions. In "Driverless Ute," Eggleton writes of a world in which waves are "strait-laced in pews," suggesting their rolling lines but also powerful forces of constraint.[64] As I suggested in the opening of this chapter, Eggleton's fantastical ute embodies connections between petroleum, cultural institutions, and climate.

As a spluttering figure, the ute also implies that satire is difficult to sustain in these conditions. A humorous engagement perhaps risks undermining climate change as a credible and urgent concern, a move aligned with corporations and lobbying groups that support climate denial. The New Zealand government, for its part, dragging its heels over emissions reductions, has been accused repeatedly of "not taking the issue seriously."[65] As I have suggested in reading *Erewhon*, however, satirical humor does not so much respond to as give form to serious critical and imaginative commitments, albeit with unstable implications because it often creates layered and indirect meanings. Eggleton's story of satire's decline perhaps also speaks to the relative marginality of humor in contemporary New Zealand European environmental culture, where the somber, nationalist aesthetic of writers like Baughan and Reeves arguably exerts a lingering force. Geoff Park describes Reeves's poem "The Passing of the Forest" as having become "an anthem for the New Zealand conservation movement."[66] *Erewhon*, relying on satire, humor, uncertainty, and wariness to illuminate the violence and contingency of settler-capitalist relationships with human and nonhuman beings, is an alternative to forms of writing that lament ecological loss and insist on authority while avoiding a critical engagement with power dynamics and economy.

The deeply muted figure of satire offered in Eggleton's "Driverless Ute," placed alongside the unchecked vibrancy of *Erewhon's* satirical humor, might be understood in the context of Eggleton's awareness of profound constraint and even of the inevitability of continued climate-related upheaval, an awareness that was unavailable to Butler. This difference could imply that critically addressing violence and injustice has become unthinkable in a time of awareness of climate change. And yet, *Erewhon's* imaginative alterity, and particularly its exuberant satirical insistence on critique and on calling up possibilities, might rather illuminate such insistence as a persistent, if more subdued, concern of contemporary environmental culture. In particular, Butler's use of satire to open spaces of critique and hope turns me toward the possibility that the satire in Eggleton's "Driverless Ute" might evoke futures that include but also exceed its final images of endings and continued loss. In his imaginary of satire's decline, Eggleton highlights not the impossibility of change but rather our unwillingness to face satire's fiery critiques and imaginaries of potential—concerns so evident in *Erewhon*—as we cling to modern forms of life involving oil and face futures involving severe ecological upheaval. The ute struggles to go on amid ecological depletion and constraint, a plaintive image that hints at grief for and attachment to the very modern institutions that Eggleton satirizes. As Stephanie LeMenager writes, describing the melancholic relationship with oil evident in contemporary United States archives, "Loving oil to the extent that we have done in the twentieth century sets up the conditions of grief as conventional oil resources dwindle."[67]

In "Driverless Ute," affective attachment to oil-intensive forms of life, such as driving, collides with a literary form that blasts such attachment as absurd, contingent, aggressive, and dangerous, so as to suggest that things could be different. In contrast with the relatively solitary journey of our narrator and Chowbok in *Erewhon*, various things, from the farm to hills, Māoritanga, and the town, are "bundled" on Eggleton's ute.[68] The extremity in this image of things casually thrown together implies that tensions around potential futures, including Māori struggles for justice in relation to lands and waters appropriated for farming and other uses, are elided in efforts to continue as usual. But Eggleton's ute is also disturbing for what it does not carry. Cities, the places where most people live, are missing. (In an earlier poem in the same collection, "Time of the Icebergs," we find Dunedin, a coastal city affected by winter cold but also vulnerable to seemingly distant upheavals to ice, the "chill fingerbones that touch you from far away, / in the time of the icebergs."[69] As the ocean rises, in part through the disintegration of ice, it is expected to drive changes in Dunedin's ground water levels, causing inundation of large expanses of the city.[70]) Eggleton suggests that many will fall away from efforts to maintain life as usual, emphasizing the need for structural transformation, and particularly isolating oil and agriculture as among the forces shaping climate change. The poem ends with a somber image of continued ecological disintegration, likening the struggling ute to empty water tanks at the end of summer, "leaving the taste of grinding/peppermill dust and dry forest floor."[71] The poem's satirical humor suggests that there is room to maneuver in relation to such a future. As Haugerud writes, it is up to the reader to decide whether satire's claims on such possibilities are "simply a joke."[72]

Note

1. David Eggleton, "Driverless Ute," in *Time of the Icebergs: Poems* (Dunedin: Otago University Press, 2010), line 9.
2. Ibid.
3. Number eight wire refers to fencing wire that is associated with ingenuity because it has been put to varied, creative uses in New Zealand.
4. Eggleton, "Driverless Ute," 9.
5. Ibid.
6. Samuel Butler, *Erewhon*, ed. Peter Mudford (London: Penguin, 1970), 255.
7. Peter Mudford, introduction to *Erewhon*, by Samuel Butler (London: Penguin, 1970), 8–9; Sue Zemka, "*Erewhon* and the End of Utopian Humanism," *ELH* 69, no. 2 (Summer 2002): 439.
8. Robert C. Elliott, *The Power of Satire: Magic, Ritual, Art* (Princeton: Princeton University Press, 1960), 101.
9. Angelique Haugerud, *No Billionaire Left Behind: Satirical Activism in America* (Stanford: Stanford University Press, 2013), 10.
10. Butler, *Erewhon*, 4.
11. On connections between satire and utopia, see Robert C. Elliott, *The Shape of Utopia: Studies in a Literary Genre* (Chicago: University of Chicago Press, 1970), 3–24.
12. Fredric Jameson, *Archaeologies of the Future: The Desire Called Utopia and Other Science Fictions* (London: Verso, 2005), 22.
13. Ibid., 23.
14. Elliott, *The Power of Satire*, 14.
15. Ibid., 33.
16. Elliott, *The Shape of Utopia*, 24.
17. Benjamin Morgan, "How We Might Live: Utopian Ecology in William Morris and Samuel Butler," in *Ecological Form: System and Aesthetics in the Age of Empire*, ed. Philip Steer and Nathan K. Hensley, 139–160.
18. Elliott, *The Shape of Utopia*, 101.
19. Jameson, *Archaeologies*, 23; Butler, *Erewhon*, 255.
20. Samuel Butler, preface to *Erewhon* (London: Penguin, 1970), 33.
21. New Zealand Government. "Statutory Acknowledgement for the Rangitata River," in *Ngāi Tahu Claims Settlement Act* (1998), Schedule 55, 344.

22. Butler, *Erewhon*, 49.

23. For an account of the ecological upheaval of New Zealand floodplains, see Geoff Park, *Ngā Uruora: The Groves of Life: Ecology and History in a New Zealand Landscape* (Wellington: Victoria University Press, 1995).

24. Eric Pawson and Peter Holland, "Lowland Canterbury Landscapes in the Making," *New Zealand Geographer* 61 (2005): 167–75.

25. Park, *Ngā Uruora*, 36.

26. Department of Conservation and Katie Williams, *Native Plant Communities of the Canterbury Plains* (Christchurch: Department of Conservation, 2005), 7.

27. Anne Glenny Wilson, "A Spring Afternoon in New Zealand," in *The New Place: The Poetry of Settlement in New Zealand, 1852–1914*, ed. Harvey McQueen (Wellington: Victoria University Press, 1993), 75.

28. Blanche Baughan, "A Bush Section," in *The New Place: The Poetry of Settlement in New Zealand, 1852–1914*, ed. Harvey McQueen (Wellington: Victoria University Press, 1993), 196.

29. Ibid.

30. Ibid.

31. Ibid., 100.

32. Quoted in Geoff Park, "Swamps Which Might Doubtless Easily Be Drained," in *Theatre Country: Essays on Landscape & Whenua* (Wellington: Victoria University Press, 2006), 187.

33. Ibid.

34. William Pember Reeves, "The Passing of the Forest," in *The New Place: The Poetry of Settlement in New Zealand, 1852–1914*, ed. Harvey McQueen (Wellington: Victoria University Press, 1993), 111.

35. Philip Steer, "Colonial Ecologies: Guthrie-Smith's *Tutira* and Writing in a Settled Environment," in *A History of New Zealand Literature*, ed. Mark Williams (Cambridge: Cambridge University Press, 2016), 88.

36. Reeves, "The Passing," 112.

37. Ibid.

38. Geoff Park, "A Moment for Landscape," in *Theatre Country: Essays on Landscape & Whenua* (Wellington: Victoria University Press, 2006), 199.

39. On this history, see Park, "A Moment," 202.

40. Ibid., 204.

41. Reeves, "The Passing," 113.

42. Butler, *Erewhon*, 46.

43. Elliott, *The Power of Satire*, 190.

44. Haugerud, *No Billionaire*, 8.

45. For a discussion of how media and commodities have linked New Zealand identity to farming and to associated life forms and objects, such as

sheep, clothing, and fencing wire, see Fiona Barker, "New Zealand Identity—Symbols of Identity," *Te Ara—the Encyclopedia of New Zealand*, http://www.TeAra.govt.nz/en/new-zealand-identity/page-7.

46. Haugerud, *No Billionaire*, 32.
47. Butler, *Erewhon*, 255.
48. Haugerud, *No Billionaire*, 32.
49. Ibid.
50. Butler, *Erewhon*, 39.
51. William Pember Reeves, "A Colonist in His Garden," in *An Anthology of New Zealand Poetry in English*, ed. Jenny Bornholdt, Gregory O'Brien, and Mark Williams (Auckland: Oxford University Press, 1997), 497.
52. Butler, *Erewhon*, 55.
53. Haugerud, *No Billionaire*, 10.
54. Butler, *Erewhon*, 45.
55. Ibid., 53.
56. Ibid., 54.
57. Ibid., 96.
58. Ibid., 255.
59. Haugerud, *No Billionaire*, 31.
60. Apirana Taylor, "Sad Joke on a Marae," in *An Anthology of New Zealand Poetry in English*, ed. Jenny Bornholdt, Gregory O'Brien, and Mark Williams (Auckland: Oxford University Press, 1997), 99.
61. Alice Te Punga Somerville, *Once Were Pacific: Māori Connections to Oceania* (Minneapolis: University of Minnesota Press, 2012), 40.
62. Taylor, "Sad Joke on a Marae," 99.
63. Butler, *Erewhon*, 257–58.
64. Eggleton, "Driverless Ute," 9.
65. Frances Cook, "Govt not taking climate change seriously—Labour," *Newstalk ZB*, April 20, 2016.
66. Park, *Ngā Uruora*, 199.
67. Stephanie LeMenager, *Living Oil: Petroleum Culture in the American Century* (Oxford: Oxford University Press, 2014), 102.
68. Eggleton, "Driverless Ute," 13.
69. David Eggleton, "Time of the Icebergs," in *Time of the Icebergs: Poems* (Dunedin: Otago University Press, 2010), 6.
70. Timothy Brown, "Effect of Sea Level Rise on Dunedin Revealed," *Otago Daily Times*, July 19, 2016.
71. Eggleton, "Driverless Ute," 17–18.
72. Haugerud, *No Billionaire*, 203.

AFTERWORD

They Would Have Ended by Burning Their Own Globe

Karen Pinkus

> To Mr. F. R. Starr, Engineer, 30 Canongate, Edinburgh.
> IF Mr. James Starr will come to-morrow to the Aberfoyle coal-mines, Dochart pit, Yarrow shaft, a communication of an interesting nature will be made to him . . .
>
> —JULES VERNE, *The Underground City, or, The Child of the Cavern*

To be sure I—a middle-aged female academic, writing in the era of 400+ ppm of CO_2, today, under a regime that fills me with fear and rage—am far from the intended recipient of the missive that opens Jules Verne's 1877 novel, *Les indes noires* (*The Black Indies*, also known as *The Child of the Cavern* or *The Underground City*). Still, the conventions of the novel address me, calling me to take my place, to prepare for a fictive trip. In theory, I can access and lose myself in this narrative as easily as a young boy reading it in French, serialized in *Le temps* or in illustrated, bound book format a short while later; or in an English translation, in the late nineteenth century.[1] I endeavor to imagine I am Starr, who has a woman to pack his bags and make sure he is well fed before he embarks on his journey and to keep his lodgings in fine order while he is gone. He sends a message to his gentlemen's club to inform fellow members that he will be absent. This detail will be important later in the novel because it will help spur the recovery effort when he goes missing. But—SPOILER ALERT—have no fear—it will all turn out fine because a mysterious beneficent force in the form of a fairy or sprite (a "brownie," eventually unmasked as an innocent wild-child, Nell) will provide Starr and his companions, the Fords, with food and water when they are trapped. But more important for me, as a reader: Before I knew what I know now, in my own past, I would have expected I would come back to this club, sit near a warm fire as Starr told his story to other men. While I have lived with some realization of my difference from such interlocutors—those gentlemen who demand to hear and so also justify the narrative—I have also come to expect, from fictions of this sort, a *nostos* or at least a minimal degree of comforting continuation and now, today, that is no longer possible for me, unless reading is nothing short of a deliberate leap into false consciousness. How can I—how can we—read this text, today?

> [Starr] belonged to an old Edinburgh family, and was one of its most distinguished members. His labours did credit to the body of engineers who are gradually devouring the carboniferous subsoil of the United Kingdom, as much at Cardiff and Newcastle, as in the southern counties of Scotland.[2]

Starr is one those Vernian engineers who can solve any problem you put before him. Lacking an immediate problem, they will invent one. They are usually bachelors, not weighed down by a needy wife or family. They are usually not industrialists because they are not driven by motives of profit but the quest for scientific domination over nature. Though most of Verne's protagonists seem comfortable, they don't work to amass wealth. They simply have it.

> We know that the English have given to their vast extent of coal-mines a very significant name. They very justly call them the "Black Indies," and these Indies have contributed perhaps even more than the Eastern Indies to swell the surprising wealth of the United Kingdom.[3]

Verne did not coin "Black Indies" to refer to the coal mine. He read it and repeated it. In fact, while he did visit Scotland several times he never himself went below ground.[4] He had guidebooks, geological treatises, and literary texts (the writings of Sir Walter Scott, among others) at his disposal. The English have named—and in Verne naming is the first act of "good colonization"—a vast subterranean realm. Like the other "good colonies" in the author's tales, there are no dark natives to displace. Blackness is a signifier of the realm but I will never have to experience it during my reading. It drops away. So Starr leaves and I leave, but the journey is interrupted abruptly, by a scientific voice.

> The better to understand this narrative, it will be as well to hear a few words on the origin of coal. . . . When James Starr had retired [from

his position at Aberfoyle several years prior], it was with the full conviction that even the smallest vein had been completely exhausted.[5]

What comes in the space of the dots above, however, are not *a few words*, but thousands, on the origins and evolution of coal in geological time. I feel compelled to consider the space or the time of hearing taken up by this lecture, recited by Verne but with the alibi of Starr himself to verify the facts. It disrupts the narrative and corresponds to the passage of time that the reader/Starr spends on the trip from Edinburgh to the mine. This excursus into geological nonfiction is far longer than any single chapter in the novel. By the time I get to the end I've lost my bearings. I have forgotten the narrative that preceded it and, in a hallucinatory state, I am not prepared for any particular events to follow. Certainly, this elucidation of coal is in line with the didactic and financial aims of Pierre Jules Hetzel, Verne's editor/father-figure. It is the kind of flat, factual prose Hetzel encouraged, in general, even if he had issues with drafts of this novel in other regards. The geo-knowledge, taken at times almost verbatim from secondary sources, is not too far removed from what we know of coal today, but with one huge exception: Verne does not know—he could not be expected to know—about the effects of accelerated historical greenhouse gas emissions on planetary warming. He therefore makes no mention of what will happen once carbon dioxide from the burning of coal will have crossed a certain threshold of concentration. Then again, he makes no mention of what happens to coal once it is removed from the ground either. This is not an industrial novel: Combustion is deferred, displaced.

There is a looming crisis in *Les indes*, however: scarcity. To be sure, Verne was optimistic that another source of fuel would eventually replace coal, so the crisis is not truly an existential disruption of the highest order. This is not to say that *Les indes noires* is not dark. The novel subscribes to a prevailing theory of the death of the sun (and consequent death of earth), but imagines that this world-ending catastrophe will take place only in a remote future. A sense of melancholy creeps in from time to time, and alarming characters (a mean-tempered old man and a snowy owl, for instance) try to thwart the progress of the miners. Is it possible that even scarcity is painfully unreadable today because it cannot come soon enough . . . or rather, it has not come soon enough.

Coal subtends the Victorian novel, perhaps invisible but powerfully embedded in language—as Hensley and Steer argue in their contribution to this collection. In Verne, its deep history is recounted on the very surface, almost as if to foreclose the possibility of its unconscious, or better, subconscious effects. Verne seems to tell his readers not to worry about what goes on below, because it's all happening here and now, and if there is a below it's just a mirror of the surface anyway. If the un/sub-conscious has already been colonized (in the positive Vernian sense of this word as the triumph of science) it is no longer a threat to rise up and intrude on the/ my surface. Yet the sub/un-conscious by definition is that which resists, so I am left in a double bind. Right here and right now. I can follow the narrative, but I know something about coal that Verne-Starr doesn't. I can't un-know what I know as much as I dream of going back and of forgetting. I can't read unless I forget, but if I forget I am reading in bad faith.

Back in the flow of narrative, Starr meets up with Ford, the old manager who had summoned him to Aberfoyle. He still lives underground in a "cottage," even though the mine has long been considered spent, along with his wife, his son, Harry, and the mine/wetnurse (*nourrice*) herself.[6] Ford has no intention of breaking up this bizarre love quadrangle. No thought of a wife for Harry ("And who would it be? A girl from up yonder, who would love merry-makings and dancing, who would prefer her clan to our mine! Harry wouldn't do it!") so no danger that this novel will veer toward a marriage plot. The Fords don't work (anymore), but all of their wants are provided for, including illumination, a fully balanced diet, and aged scotch. They have no desire to go up to the surface (where the coal is burned, creating the sooty industrial landscape that they abhor). Off they go, on a mission to discover the new seam, but they are trapped! And then Verne cuts his own seam, simply inserting a flash forward:

> 3 years later . . .
> Arrived in Coal Town, the visitor found himself in a place where electricity played a principal part as an agent of heat and light. Although the ventilation shafts were numerous, they were not sufficient to admit much daylight into New Aberfoyle, yet it had abundance of light. This was shed from numbers of electric discs; some suspended from the vaulted roofs, others hanging on the natural pillars—all, whether suns or stars in size, were fed by continuous currents produced from electro-magnetic machines.[7]

We skip the labor and violence that led to the opening of a new source of coal. We skip any discussion of investors or industrialists who finance the operation. We are spared the stages in between the building of the city as we skip the science. Electricity just happens, just as capital just happens. In fact, in Verne electricity is essentially "free" energy without labor.[8] Community happens. Verne uses the term *"grande famille"* throughout this text to refer to the underground dwellers at Aberfoyle. The coal town mirrors the surface, but without any of the social or environmental problems of Dickens's Coketown. However, as it turns out, in *Les indes, la grande famille* is a relatively new development. We learn of Ford's ancestors:

> They labored like convicts at the work of extracting the precious combustible. It is even believed that the coal miners, like the salt-makers of that period, were actual slaves.[9]

But now the workers are happy and fully invested in the labor they do, a fantasy that seems inextricable from that of coal itself. Yet—and in spite of Hetzel's appeals to the author to the contrary—Verne does not include any descriptions of the process. Instead, we find the city fully functioning, while colliers stroll past underground lakes and parks, workers at leisure.[10]

I follow young Harry Ford and Starr discussing the extraordinary qualities of coal:

> "Indeed," cried the young man, "it's a pity that all the globe was not made of coal; then there would have been enough to last millions of years!"
> "No doubt there would, Harry; it must be acknowledged, however, that nature has shown more forethought by forming our sphere principally of sandstone, limestone, and granite, which fire cannot consume."
> "Do you mean to say, Mr. Starr, that mankind would have ended by burning their own globe?"
> "Yes! The whole of it, my lad," answered the engineer. "The earth would have passed to the last bit into the furnaces of engines, machines, steamers, gas factories; certainly, that would have been the end of our world one fine day!"[11]

I am puzzled by the strange grammar of Harry's query and Starr's response, both in the conditional past as if the boy (the reader) could position himself in the future and look down on the networks of energy on earth at that moment and imagine an alternative geological history. Compare this with the simple future deployed by engineer Cyrus Smith in Verne's *The Mysterious Island* (1874): "One day our globe will come to an end, or rather, animal and vegetable life will no longer be possible because of an intense cooling that will take place."[12] Perhaps there is no point, in the grand scheme of things, to make too much of verb tenses in a passage that has no particular force as far as the narrative itself. Yet it seems almost as if Verne-Starr can't speak directly, constatively, about change in the past, whereas it is easier to posit global cooling in the deep future. In the conditionality of the conversation of *Les indes* I cannot unhear a certain ambivalence about coal.

> In my opinion England is very wrong in exchanging her fuel for the gold of other nations![13]

This protectionist nationalism is totally predictable in Vernian terms. The resources of the subsurface belong to and should stay on the surface of the nation-state, shouldn't they? There is a certain logic to Starr's statement and I can only oppose it using an argument that would take place outside of the bounds of the novel, so I would have to stop reading. Or perhaps read a different kind of prose?

> We will end the war on coal, and rescind the coal mining lease moratorium, the excessive Interior Department stream rule, and conduct a top-down review of all anti-coal regulations issued by the Obama Administration. We will eliminate the highly invasive "Waters of the US" rule, and scrap the $5 trillion dollar Obama-Clinton Climate Action Plan and the Clean Power Plan and prevent these unilateral plans from increasing monthly electric bills by double-digits without any measurable effect on Earth's climate. Energy is the lifeblood of modern society.[14]

So painful, I have to turn away. I could retreat to the supernatural, right there on the pages of *Les indes*:

> "A goblin, a brownie, a fairy's child," repeated Jack Ryan, "a cousin of the Fire-Maidens, an Urisk, whatever you like! It's not the less certain that without it we should never have found our way into the gallery, from which you could not get out."[15]

Surface dweller and bag-piper Jack Ryan continues to believe in magical forces, even thought Verne offers us a real explanation for the rescue of the Fords and Starr: the good will of the foundling, Nell, and the giant subterranean-dwelling snowy owl that protects her. The father-editor Hetzel begged Verne to tone down the "Barnum" style of the novel's ending: fires, geological upheavals, the revenge of Nell's grandfather, the evil Silfax (Lucifer). First let the book have success, the editor says. Then you can write a crazy epilogue. In the end, the product we have is the result of a compromise, the text was rewritten at least four times, perhaps more.

Someday the new seam will be spent but Harry and Nell will continue the line just as the subsurface will continue to offer refuge (perhaps even serving as camps for Britain's indigent classes, Verne suggests!).

La grande famille. We must take care of each other, but can we imagine doing so outside of this escapist fantasy? Is this the same kind of fantasy that comes from or is described by "resilience"? How can I—how can we—read today, knowing what we know and unable to unknow about massive global disruption? Sure, Jules Verne had his share of problems on the surface of the land, according to biographies. Accused of plagiarism, he was tormented by lawyers. He suffered poor health. He was estranged from his son. He desired to be named to the French academy and was frustrated that he was not considered a serious author. One has to adapt.

> "Resilience" refers to the ability to anticipate, prepare for, and adapt to changing conditions and to withstand, respond to, and recover rapidly from disruptions.[16]

What does it mean to live and read through the lens of climate change? How I would like to get outside, to think otherwise. Some days it's possible, but mostly, not. I'm not sure that we can now absorb narratives of a certain kind of progress, even if non-linear and dark. Narratives that think forward in time with a certain kind of openness to a future that is enough like the present to provide some sense of stability and yet different enough from the present to provide the impetus to keep reading . . .

> "Fragility" refers to a condition that results from a dysfunctional relationship between state and society and the extent to which that relationship fails to produce policy outcomes that are considered effective or legitimate.[17]

Notes

1. Hetzel felt that this particular work was better suited for *Le temps* rather than the *Magasin d'Éducation et de Récréation* where he serialized most of Verne's novels because he hoped *Les indes* would be less an adventure fantasy and more a depiction of social reality of the coal mine. See Olivier Dumas, Piero Gondolodella Riva, and Volker Dehs, eds. *Correspondance inédite de Jules Verne et de Pierre-Jules Hetzel (1863–1886)*, vol. II (1875–1878) (Geneva: Slatkine, 2001), 123.

2. Jules Verne, *The Underground City, or, The Child of the Cavern*, trans. William Henry Kingston (Philadelphia: Porter Coates, 1883), 2.

3. Ibid., 3.

4. Verne also visited the mines at Anzin, but unlike Zola, who researched his coal epic *Germinal* there, Verne never went down below. For an account of the different experiences and commitments, see Henri Marel, "Jules Verne, Zola e la mine," *Les Cahiers naturalistes* 54 (1980): 197–200.

5. Verne, *Underground*, 20, 28.

6. Let us not forget that in industrial England, the cottage is the space of the hand loom that persists for decades alongside the factory. "Cottage" is, in a sense, the anti-factory.

7. Verne, *Underground*, 132.

8. Jean Chesneaux, *Une lecture politique de Jules Verne* (Paris: François Maspero, 1971), 40.

9. Verne, *Underground*, 49.

10. Actual (gold) mining, as Verne noted elsewhere, is debasing. Jean Chesneaux, *Jules Verne. Un regard sur le monde* (Paris: Bayard Editions, 2001), 218. Catherine Gallagher notes that early nineteenth-century reformers deployed an analogy of factory workers to slaves as a means of emphasizing the repressively long hours, lack of protection, and physical brutality of the factory system. See her *Industrial Reformation of English Fiction, 1832–1867* (Chicago: University of Chicago Press, 1980), 11.

11. Verne, *Underground*, 34.

12. Jules Verne, *The Mysterious Island*, trans. Jordan Stump (New York: Random House, 2001), 231.

13. Verne, *Underground*, 35.

14. Donald Trump, "Energy Independence," https://greatagain.gov/energy-independence-69767de8166#.v8909aqfx. Accessed December 30, 2016.

15. Verne, *Underground*, 137.

16. Barack Obama, Presidential Memorandum on Climate Change and National Security, September 21, 2016, https://www.whitehouse.gov/the-press-office/2016/09/21/presidential-memorandum-climate-change-and-national-security.

17. Ibid.

ACKNOWLEDGMENTS

This is a book about global systems and collaboration so it feels appropriate that the ensemble of actors that brought it into being is vast and multiply scaled. This network of connections joins Australasia to North America, and links into the same storyline a cast of institutions, people, and material agencies too numerous to name—including Google Docs, Skype, and the energy running those programs. The initial kernel of this project was a pair of connected panels at the North American Victorian Studies Association in Honolulu, Hawaii in 2015. We thank the organizers of that conference, the contributors who joined us on those panels—Sukanya Banerjee, Jen Hill, Liz Miller, and Lynn Voskuil—and the engaged audiences we found there. The planning and editing of the book occurred through digital channels connecting an office in Palmerston North, New Zealand (UTC +12:00), to a basement in Silver Spring, Maryland (UTC −5:00). For materially supporting that work, we thank the Graduate School and the Lafferty Endowed Fund for English at Georgetown University and the College of Humanities and Social Sciences at Massey University.

People are what matter. The project was brought to life by the patient and inspiring collaborators whose bylines appear in the foregoing pages that follow. We also thank Kathleen Frederickson, Devin Griffiths, Barbara Leckie, Tobias Menely, Zach Samalin, and the Eighteenth- and Nineteenth-Century Atlantic Cultures Workshop at the University of Chicago. The students of 139.307, "Victorian Literature," and English 145, "The Nineteenth Century British Novel," have helped us come to terms with our own arguments. Our editor, Richard Morrison, supported the vision for this project from its earliest incarnation. Daniel Worden and Elaine Freedgood believed in the book and provided inspiration for it. For acute and indispensable editorial help, we thank John James, Kirsten Ellmers, and the staff at Fordham University Press. Much earlier and abridged versions of chapters by Sukanya Banerjee and Jesse Oak Taylor originally appeared in *Victorian Studies* 58.2 (Winter 2016), and are reproduced with permission from Indiana University Press. All errors belong only to us.

Finally, and most important, our indescribable gratitude goes to our families: Sarah, Esther, Silas, Joseph, and Amos; Anne, June, and Irene. We love you; you are us.

CONTRIBUTORS

MONIQUE ALLEWAERT is Associate Professor of English at the University of Wisconsin–Madison. She is the author of *Ariel's Ecology: Personhood and Colonialism in the American Tropics, 1760–1820* (2013).

SUKANYA BANERJEE is Associate Professor of English at the University of Wisconsin-Milwaukee. She is the author of *Becoming Imperial Citizens: Indians in the Late-Victorian Empire* (2010) and coeditor of *New Routes for Diaspora Studies* (2012).

ADAM GRENER is Lecturer in English at Victoria University of Wellington. His current book project is "Improbable Realism: Chance, the Rise of Statistics, and the Nineteenth-Century British Novel."

NATHAN K. HENSLEY is Associate Professor of English at Georgetown University. He is the author of *Forms of Empire: The Poetics of Victorian Sovereignty* (2016).

DEANNA K. KREISEL is Associate Professor of English at the University of British Columbia. She is the author of *Economic Woman: Demand, Gender, and Narrative Closure in Eliot and Hardy* (2012).

ELIZABETH CAROLYN MILLER is Professor of English at the University of California, Davis. She is the author of *Slow Print: Literary Radicalism and Late Victorian Print Culture* (2013) and *Framed: The New Woman Criminal in British Culture at the Fin de Siècle* (2008).

BENJAMIN MORGAN is Associate Professor of English Language at the University of Chicago. He is the author of *The Outward Mind: Materialist Aesthetics in Victorian Science and Literature* (2017).

KAREN PINKUS is Professor of Italian and Comparative Literature at Cornell University. She is the author of *Fuel: A Speculative Dictionary* (2016), *Alchemical Mercury: A Theory of Ambivalence* (2009), *The Montesi Scandal: The Death of Wilma Montesi and the Birth of the Paparazzi in Fellini's Rome*

(2003), *Picturing Silence: Emblem, Language, Counter-Reformation Materiality* (1996), and *Bodily Regimes: Italian Advertising Under Fascism* (1995).

AARON ROSENBERG is a Leverhulme Early Career Fellow at King's College London. His current book project is "Scale, Modernity, and the Novel: From Realism to the Genres of Deep Time."

TERESA SHEWRY is Associate Professor of English at the University of California, Santa Barbara. She is the author of *Hope at Sea: Possible Ecologies in Oceanic Literature* (2015) and coeditor of *Environmental Criticism for the Twenty-First Century* (2011).

PHILIP STEER is Senior Lecturer in English at Massey University. His current book project is "Borders of Britishness: The Novel and Political Economy in the Victorian Settler Empire."

JESSE OAK TAYLOR is Associate Professor of English at the University of Washington. He is the author of *The Sky of Our Manufacture: The London Fog in British Fiction from Dickens to Woolf* (2016) and coeditor of *Anthropocene Reading: Literary History in Geologic Times* (2017).

LYNN VOSKUIL is Associate Professor of English at the University of Houston. She is the author of *Acting Naturally: Victorian Theatricality and Authenticity* (2004) and editor of *Nineteenth-Century Energies: Literature, Technology, Culture* (2016).

INDEX

abundance, 77, 103–6, 113–14
The Accursed Share (Bataille), 104
Ackerman, Alan R., 41n61
actor-network theory, 22–23, 33–34, 40n45
"Ad Valorem" (Ruskin), 106, 114
aesthetic value, 111–15, 184
African diaspora, 204–5, 209–10, 217
After London (Jefferies), 142, 146–47
agency. *See* shared agency and consciousness
The Age of Analogy: Science and Literature Between the Darwins (Griffiths), 5
Aguilar, Laura, 36
Alaimo, Stacy, 101, 102
Allewaert, Monique, 12, 203
Althusser, Louis, 65, 66–67, 80n12
Anderson, Katherine, 131
Anthropocene: defined, 5, 43; ecological formalism's approach to, 10–14, 42–43, 45–49, 52–59, 154–55, 162; elegy for, 45–47, 49, 52–53, 55–59; Ghosh on, 3–4; shared agency and consciousness and, 21, 93, 95–96
Anthropocene Reading: Literary History in Geologic Times (eds. Taylor and Menely), 5
anthropogenic exigency, 36–37
Arata, Stephen, 197n10
Arendt, Hannah, 77
Aristotle: *Poetics*, 23, 116–17n13
Arts of Living on a Damaged Planet: Ghosts and Monsters of the Anthropocene (Tsing, et al.), 46–47
Atlantic materialism, 203–22; colonialism and, 218–19; electric theory and, 209–18; fetishism and, 203–4, 206–10; spiritualism as key to, 203–6
An Atlas of the European Novel, 1800–1900 (Moretti), 5
atmospheric imagery, 11, 127–33
"The Attraction of Planets" (Delany), 210–12

Bakhtin, M. M., 185
Banerjee, Sukanya, 10, 21
Banks, Joseph, 166
Barrett, Ross, 66
Barthes, Roland, 144–45, 146
Bataille, Georges, 115; *The Accursed Share*, 104
Baucom, Ian, 4, 9
Baughan, Blanche: "A Bush Section," 228
Beadon (lord), 35
Beaumont, Matthew, 142
Beer, Gillian, 5, 183
Belich, James, 148–49
Bellamy, Edward: *Looking Backward*, 140, 149, 150–51, 159n51
Bengali theater, 25, 27, 31. *See also Neel Darpan* (Mitra)
Benjamin, Walter, 3
Bennett, Jane: *Vibrant Matter*, 111
Bhatia, Nandi, 28, 38n18
biblio-stratigraphy, 48–49
"The Black Diamonds of England" (Horne), 63–64
The Black Indies (Verne), 12–13, 241–48
black prometheanism, 209–10
Blake; or the Huts of America (Delany), 12, 206, 210, 211–18
Blier, Suzanne, 208
Bloch, Ernst, 148
Blythe, Helen, 158n44
The Body Economic (Gallagher), 104
Bolster, W. Jeffrey, 87, 94, 98nn8–9
Bonneuil, Christophe, 46
Bosman, William, 207
botanical scale: application to environmental change, 161–63, 178, 178n3; correlation of, 168–69; distortions of, 171–76; for global study of plants, 163–68; representation of, 169–71; sublime framework for, 176–77

253

Brady, Emily, 176, 177
Brand, Stuart, 58
Brantlinger, Patrick, 5–6, 53
Breakthrough Institute, 154
Brenneke, Ernest: *Thomas Hardy's Universe*, 194
British Empire and imperialism: background, 2–3; capitalism and, 116n2, 121–22, 133, 134–35n11, 206–8, 215, 217–19; coal as fuel for, 63–64; dramatic form and, 36–37; imperial sciences and, 162–63; interconnections with ecology, 121–33; metonymies for, 123–27; Ruskin on, 102; slavery and, 74–75, 102, 204–10, 212–13. *See also* coal, as literary effect; colonialism; systemic interconnections; utopian literature
Brown, Robert, 166
Browne, Janet, 162, 164–66
Browning, Robert: "Love among the Ruins," 13–14
Bubandt, Nils, 46–47
Buckley, Arabella, 54–55
Buller, Walter, 229
Butler, Samuel, 226; *Erewhon*, 11, 12, 139–60, 224–27, 230–38; *Erewhon Revisited*, 150; *A First Year in Canterbury Settlement*, 149–50; *Life and Habit*, 151. *See also* utopian literature

Canning (lord), 26
Cannon, Susan, 166
Capital (Marx), 64–65, 203–6, 218–19
capitalism: coal extraction and, 54–55, 64–67, 71–72, 73–78; economic surplus, 77, 103–6, 113–14; economic value, 106–7, 112–14, 115; Freud on, 141n1; imperialism and, 116n2, 121–22, 133, 134–35n11, 206–8, 215, 217–19; Marx on, 1–2, 116n2, 203–4; Nature/Society dualism and, 7–8; steam power and, 92, 95–97; utopian literature and, 143–45, 153, 155; water power and, 87–89
Capitalism in the Web of Life (Moore), 101, 103, 122
Carlyle, Thomas, 102
Carpenter, Mary Wilson, 98–99n19
catastrophism, 99n22
Çelikkol, Ayşe, 126
Chakrabarty, Dipesh, 4, 43–44, 56
Chatterjee, Bankimchandra: *Rajmohan's Wife*, 21–22
Chaudhuri, Una, 33
Chen, Mel Y., 36

Clark, Timothy, 95–96, 163, 179n10, 183, 197n7
cliffhanger, 195–96
coal, as literary effect: background, 63–64, 66–67, 242–43; in *Cranford* (Gaskell), 10, 68–71, 77; dialectical thinking and, 78–79; in escapist fantasy, 242–45; in *The Expansion of England* (Seeley), 71–73, 76; industrial metaphors for, 54–55; mode of production and, 64–66; in *North and South* (Gaskell), 10, 67–71, 76; in *Nostromo* (Conrad), 10, 73–78, 82n38. *See also* steam power; water- to steam-power transition
Cockram, Gill G., 113
Coleridge, Samuel Taylor, 109, 116–17n13
collectivization, 87–88
colonialism: Atlantic materialism and, 218–19; ecocriticism of, 24–25, 29–32, 36–37, 38n18, 41n54; as frontier of capitalism, 206; indigo cultivation and, 23–26, 28–29, 33–35; Ruskin on, 101–2, 107–8; satirical representation of, 11–12, 141; settler colonialism, 11–12, 72–73, 141–42, 147–54, 158nn44–45, 228–29, 231; sustainability and, 101–2; in utopian literature, 147–54
"A Colonist in His Garden" (Reeves), 232
"Comets" (Delany), 210–11
Conrad, Joseph, 10, 73–78, 79, 82n38
consciousness. *See* shared agency and consciousness
cosmic irony, 231
cosmic pessimism, 194
Courtin, Sébastien-Jacques, 207–8
Cowels, Henry M., 62
Craig, David M., 113
Cranford (Gaskell), 10, 68–71, 77
Crosby, Alfred, 5–6
A Crystal Age (Hudson), 142–43
crystallisation, 11, 107–10
Culture and Society (Williams), 144

Daly, Nicholas, 81n17
Darwin, Charles, 7, 141, 153, 166, 183; *On the Origin of Species*, 9, 50–51, 95, 152
Das Kapital (Marx), 64–65, 203–6, 218–19
Daston, Lorraine, 167
Davies, Jeremy, 2, 58
Debeir, Jean-Claude, 88–89
de Candolle, Augustin-Pyramus, 166
"defined excluded," 66–67
Delany, Martin, 12, 205, 210–12, 218–19, 221n25; "The Attraction of Planets,"

210–12; *Blake; or the Huts of America*, 12, 206, 210, 211–18; "Comets," 210–11
Deléage, Jean-Paul, 88–89
Deleuze, Gilles: "Immanence: A Life," 111; *A Thousand Plateaus*, 111
DeLoughrey, Elizabeth, 29, 31
Derrida, Jacques, 7
dialectical thinking, 7–8, 78–79
Dickens, Charles, 104; *Dombey and Son*, 11, 66, 121–35; *Martin Chuzzlewit*, 123; "The Wind and the Rain," 131. See also *Dombey and Son*
Dimock, Wai Chee, 43
Dombey and Son (Dickens): atmospheric imagery in, 11, 127–33; metonymy in, 123–27; railway symbolism in, 66, 124; social processes represented through, 121–23
domestic melodrama, 123–24, 134n8
domestic novel, 11
Douglass, Frederick, 205, 221n25; *The Heroic Slave*, 209–10
Dove, Heinrich, 131, 170
dramatic form: groundedness of, 10, 24–25, 36–37, 41n61; shared agency and, 22–24, 32–36, 40n45
Dreams of an English Eden (Spear), 102
"Driverless Ute" (Eggleton), 223–24, 236–38
dualism: of sail and steam ships, 75–76; of species concept, 56. See also Nature/Society dualism
Dutt, Michael Madhusudan, 27, 39n38

East India Company (EIC), 23, 25–26
ecological formalism: approach to Anthropocene, 10–14, 42–43, 45–49, 52–59, 154–55, 162; form for (see *Dombey and Son*; *The Mill on the Floss*; organicism; sustainability theory; temporal structure); future and (see Atlantic materialism; escapist fantasy; satire); method for (see coal, as literary effect; dramatic form; elegy; *In Memoriam*; *Neel Darpan*); precursor to (see Ruskin, John); scale for (see botanical scale; realist novels; utopian literature)
Ecological Imperialism: The Biological Expansion of Europe, 900–1900 (Crosby), 5–6
ecomodernism, 154–55
Ecomodernist Manifesto (Asafu-Adjaye), 154
economic surplus, 77, 103–6, 113–14
economic value, 106–7, 112–14, 115

Eggleton, David, 12; "Driverless Ute," 223–24, 236–38; "Time of the Icebergs," 237
The 18th Brumaire of Louis Bonaparte (Marx), 218–19
electric theory, 209–18
elegy: defined, 45, 51; ecocriticism and, 45–47, 51–52, 59. See also *In Memoriam* (Tennyson)
Eliot, George, 98n16; *Middlemarch*, 9; *The Mill on the Floss*, 10–11, 85–100. See also *The Mill on the Floss*
Eliot, T. S., 117n13
Elliott, Robert C., 225–26
Ellis, Erle, 154
Empson, William, 153
Endersby, Jim, 50, 61, 164, 179n11
energy systems. See coal, as literary effect; steam power; water power and water rights; water- to steam-power transition
Engels, Friedrich, 1–2, 155, 220n3
environmentalism: Ruskin on, 102–3. See also organicism; sustainability theory
environmental sublime, 177
epochal shift, 10–11, 85–86
Erewhon (Butler), 11, 12, 139–60, 224–27, 230–38
Erewhon Revisited (Butler), 150
escapist fantasy, 241–48
Essay on the Principles of Population (Malthus), 106
Esty, Jed, 184, 187
The Ethics of the Dust (Ruskin), 103, 107–10
evolutionary theory: botany and, 5; ghosts of extinction, 46–47; *In Memoriam* and, 42–44, 48–49, 52–53, 56–59; nature as agent of, 44; *On the Origin of Species*, 9, 50–51, 95, 152; scale and, 8–9, 183, 196; utopian literature and, 141–42, 151–53
excess (economic), 77, 103–6, 113–14
The Expansion of England (Seeley), 71–73, 76
extinction and extinction shifts: in *In Memoriam*, 43–44, 50, 53–54, 62n45; term usage, 43

factory literature, 67–71
Fanon, Frantz, 6
feminist utopia, 142–43
fetishism, 203–4, 206–10, 215–17
A First Year in Canterbury Settlement (Butler), 149–50
The Fixed Period (Trollope), 149
Flaubert, Gustave: *Madame Bovary*, 21

flooding: temporal structure of, 89–90, 92–93, 95–96, 98n16, 99nn21–22
Flora Indica (Hooker), 11, 162–63, 167–70, 171f
"flow" energy, 87–88
foreshadowing, 89–90, 93, 96
form. *See* ecological formalism; species concept
Forms: Whole, Rhythm, Hierarchy, Network (Levine), 5
Forster, E. M., 79
Forsyth, P. T.: "The Pessimism of Mr. Thomas Hardy," 194
Fossil Capital (Malm), 103, 155
fossil record: in *In Memoriam*, 45, 48, 54, 55–56; in *A Pair of Blue Eyes*, 195–96; species concept and, 50–51
Fourier, Robert Owen, 144
Franklin, Benjamin, 211, 212
Freedgood, Elaine, 125
Freese, Barbara, 69
Fressoz, Jean-Baptiste, 46
Freud, Sigmund, 2
"fruitful waste ground," 102–3, 108, 114
Frye, Northrop, 139, 146, 150
Fuss, Diana, 47, 59

"Gaia model" (Lovelock), 117n14
Gallagher, Catherine, 71, 104, 248n10
Galton, Francis, 168
Gan, Elaine, 46–47
Gaskell, Elizabeth, 79; *Cranford*, 10, 68–71, 77; *North and South*, 10, 67–71, 76
geological forms and record: industrial metaphors and, 54–55; in *In Memoriam* (Tennyson), 45, 48–49, 54, 55–56; in *A Pair of Blue Eyes* (Hardy), 195–96; power of, 107–10; species concept and, 50–51
Ghosh, Amitav, 3–4; *The Great Derangement: Climate Change and the Unthinkable*, 21–22, 33
Ghosh, Girish, 30–31
ghosts, of extinction, 46–47
Gidal, Eric, 48–49
Gifford, Terry, 102
Gilman, Charlotte Perkins: *Herland*, 142–43
globalization. *See* British Empire and imperialism; colonialism; systemic interconnections
Gossin, Pamela, 183, 196n4
grande famille, 245, 247
Graver, Suzanne, 97n1
gravitational theory, 211–13

Gray, Steven, 79n1
The Great Derangement: Climate Change and the Unthinkable (Ghosh), 21–22, 33
The Great Romance (anonymous), 149
Grener, Adam, 11, 121
Griffiths, Devin, 5, 47, 49
groundedness, of dramatic form, 10, 24–25, 36–37, 41n61
Guattari, Félix: *A Thousand Plateaus*, 111
Guha, Ranajit, 38n18

habits of mind and character, 151–52
Haeckel, Ernest, 9
Hallam, Arthur Henry, 43, 44, 47, 51, 57–58
Hamilton, Clive, 154–55
Hamilton, Robert, 210, 212–13
Hammond, Mary, 91
Handley, George, 29, 31
Haraway, Donna, 52
Hardy, Thomas, 182, 197n8; *A Pair of Blue Eyes*, 12, 187–88, 195–96; "The Science of Fiction," 188–89; *Tess of the D'Urbervilles*, 6–9, 13; *Two on a Tower*, 11–12, 187–94, 198n30; "Wessex Edition," 187–88. *See also* realist novels
Harrison, Robert Pogue, 46
Hartwick's Rule, 117n17
Harvey, David, 72
Haugerud, Angelique, 225, 231–32
Hay, Eloise Knapp, 76, 82n34
Heise, Ursula, 45
Hémery, Daniel, 88–89
Hensley, Nathan K., 10, 24, 63, 85, 89, 99n21
Herland (Gilman), 142–43
The Heroic Slave (Douglass), 209–10
Hetzel, Pierre Jules, 243, 247n1
Hickman, Jared, 209–10
Hill, Jen, 168
Himalayan Journals (Hooker), 11, 162–63, 164, 166–68, 170–77
Hobsbawm, E. J., 80n8
Hobson, J. A., 106
Hooker, Joseph Dalton, 162–63; *Flora Indica*, 11, 162–63, 167–70, 171f; *Himalayan Journals*, 11, 162–63, 164, 166–68, 170–77; *On the Distribution of Heat over the Surface of the Globe*, 168; *The Rhododendrons of the Sikkim-Himalaya*, 164, 165f. *See also* botanical scale
hope: escapist fantasy and, 241–48; realist novels and, 195–96; through Atlantic materialism, 203–22 (*see also* Atlantic

Index

materialism); through satire, 223–40 (see also satire)
Horne, Richard H., 63–64
Howard's End (Forster), 79
"How We Live and How We Might Live" (Morris), 143–46, 153
Hudson, W. H.: *A Crystal Age*, 142–43
humanism, 56
Humboldt, Alexander von, 117nn13,14, 166; *Personal Narrative of Travels to the Equinoctial Regions of the New Continent, 1799–1804*, 172
humor, 224–30, 232–34, 235f, 236–37
Huxley, T. H., 43, 54, 55

"Immanence: A Life" (Deleuze), 111
immortality, 44–46, 51–53
imperialism. *See* British Empire and imperialism
indexes, 208–9
indigo cultivation and trade: background, 23, 35–36, 38n24; colonialism and, 23–26, 28–29, 32–35
The Indigo Planting Mirror (translation), 27, 39nn27,38
indigo rebellion, 26–27
industrialism: elegy and, 54–56; escapist fantasy and, 244; Marx on, 203–4; romance novel and, 67–71; Ruskin on pollution from, 102; satire and, 231, 236; social transition in, 97n1; utopian literature and, 140–44; water- to steam-power transition, 10–11, 85–86, 91–92, 95–97, 98n8
industrial metaphors, 54–56
The Industrial Reformation of English Fiction (Gallagher), 71
Industrial Revolution, 1–2, 4, 8
In Memoriam (Tennyson), 42–62; as Anthropocene elegy, 45–47, 49, 52–53, 55–59; background, 43; as commemoration, 44, 46, 59; context for, 10, 42–43; extinction shifts in, 43–44, 50, 53–54, 62n45; immortality in, 44–46, 51–53; industrial metaphors in, 54–56; mourning for species in, 10, 44–48, 57, 59; nature as agent of evolution in, 44, 45–46; shared agency in, 51–52, 55–58; species concept in, 44, 47–48, 50–51, 56–57; structure of, 47–49, 51, 52–53
international capitalism, 116n2, 121–22, 133, 134–35n11, 206–8, 215, 217–19
iron, 11, 112, 115
irony, 77–78, 225, 230–36
iterative stanzas, 48

Jaffe, Audrey, 125
Jahan Ramazani, 45
James, Henry, 99n21
Jameson, Fredric, 64–65, 78–79, 81n18, 126, 140, 145, 146, 186, 225
jatra, 30–31, 34, 39n38, 40n52
Jefferies, Richard: *After London*, 142, 146–47
Jevons, William Stanley, 143
Johnston, F. W., 104

Kershaw, Baz, 23
Klotz, Michael, 128
Kohn, Eduardo, 46
Kreisel, Deanna K., 11, 86, 101
Kulinkulsharbashwa (Tarakratna), 27

laissez-faire economic policy, 105–6
Lang, Andrew: "Realism and Romance," 186–87
large-scale systems, 144–45
Larsen, Anne, 164
Latour, Bruno, 11, 22–23, 36, 41n54, 134n5, 163
Law, Jules, 87, 90
LeMenager, Stephanie, 237
Lenin, Vladimir, 116n2
Les indes noires (Verne), 12–13, 241–48
Levin, Simon, 162
Levine, Caroline, 5, 50, 113, 115, 177
Levine, George, 110–11, 186, 193–94
Lévi-Strauss, Claude, 7
Lewis, Simon, 52–53
life, Ruskin's definition of, 11, 106–12, 115, 118n40
Life and Habit (Butler), 151
Lloyd, Trevor, 235f
London, overdevelopment of, 182
Long, James, 27, 39n26
Looking Backward (Bellamy), 140, 149, 150–51, 159n51
"Love among the Ruins" (Browning), 13–14
Lovelock, James: "Gaia model," 117n14
Luciano, Dana, 36
Lukács, Georg, 7–8, 185
Lyell, Charles, 45, 100; *Principles of Geology*, 43, 53, 119n55

MacDuffie, Allen, 5, 102–3, 106, 114, 121
Madame Bovary (Flaubert), 21
Makandal, François, 207–8, 209, 216
Malm, Andreas, 70, 86, 87–88, 96–97, 103; *Fossil Capital*, 103, 155

Malthus, Thomas Robert, 117n20, 142–43; *Essay on the Principles of Population*, 106; *Principles of Political Economy*, 106
Marcus, Steven, 123–24
marginal utility theory, 117n20
Martin Chuzzlewit (Dickens), 123
Marx, Karl: *The 18th Brumaire of Louis Bonaparte*, 218–19; *Capital*, 64–65, 203–6, 218–19; on capitalism, 1–2, 116n2, 203–4; on fetishism, 12, 203–4; on imperial expansion, 116n2; on mesmerism, 220n3; on modernity, 1–2, 206; as proto-ecologist, 141; utopian literature and, 153, 155
Marx, Leo, 140
Maslin, Mark, 52–53
mass extinction. *See* extinction and extinction shifts
materialism, 110–11, 115. *See also* Atlantic materialism
Mattes, Eleanor Bustin, 45
Maynard, Jessica, 115
McCauley, Alex, 50
McGurl, Mark, 185
mechanical form, 109
Menely, Tobias, 35
Mershon, Ella, 109–10
Mesmer, F. A., 209
mesmerism, 204–5, 209, 220n3
metals, vitality of, 11, 111–12, 115
metonymies, 123–27
Middlemarch (Eliot), 9
Mill, John Stuart: *Principles of Political Economy*, 106
Miller, Elizabeth Carolyn, 10, 85
Miller, John MacNeill, 51–52
The Mill on the Floss (Eliot), 85–100; foreshadowing in, 89–90, 93, 96; perspective in, 93–95; plot, 86, 87, 89, 100n36; review of, 87; steam-power in, 89, 91–93, 95, 98n8; time scale in, 95–96; transition between water- and steam-power, 10–11, 85–86, 91–92, 95–97, 98n8; water power and water rights in, 86–89, 95, 99nn21–22. *See also* temporal structure
Mitra, Dinabandhu, 25–26; *Neel Darpan*, 22–32. See also *Neel Darpan*
mode of production, 64–65
modernity: climate change and, 3–8, 16n22; coal and, 63–65, 76–78; Marx on, 1–2, 206; Nature/Society dualism and, 95–96; pace of life, 91–93; utopian literature and, 154–55

Modern Painters (Ruskin), 108
modern Realists, 186–87
Moore, Jason: on capitalism, 101, 103, 114, 155; *Capitalism in the Web of Life*, 101, 103, 122; on colonialism, 41n54, 206; on modernity, 6; on Nature/Society dualism, 7, 24, 34, 56, 96, 122, 128; on organicism, 120n74
Morant Bay Rebellion (1865), 102
More, Thomas: *Utopia*, 225
Moretti, Franco, 5, 178; *An Atlas of the European Novel, 1800–1900*, 5
Morgan, Benjamin, 11, 94, 100n34, 139, 167, 226
Morley, Henry: "The Wind and the Rain," 131
Morris, William: "How We Live and How We Might Live," 143–46, 153; *News from Nowhere*, 11, 140–60. *See also* utopian literature
Morse, Samuel, 211
The Mortal Sea: Fishing the Atlantic in the Age of Sail (Bolster), 87
Morton, Timothy, 32, 45, 53–54, 178n3
mourning, 10, 44–48, 57, 59. See also *In Memoriam*
The Mysterious Island (Verne), 245–46

nationalism, 24–25, 71–73, 76, 246
nature: as agent of evolution, 44, 45–46; human assertion of normalcy on, 94
Nature/Society dualism: capitalism and, 7–8; in *The Mill on the Floss*, 96; modernity and, 95–96; Moore on, 7, 24, 34, 56, 96, 122, 128; organicism and, 118n13; utopian literature and, 11, 141–44, 146–48, 150–54
Neel Darpan (Mitra), 22–32; anticolonialism and, 24–25, 27, 28–29, 31, 38n18; background, 25–26; dialogue, 20, 28–31, 34; ecocriticism and, 24–25, 29–32, 36–37, 38n18; groundedness of, 10, 24, 36–37, 41n61; impact of, 31; indigo's role in, 25, 32–36; plot, 23, 27–28, 32–33; *Rajmohan's Wife* comparison, 22; rape scene, 28–29; review of, 36
Nemesvari, Richard, 189, 198n28
News from Nowhere (Morris), 11, 140–60. *See also* utopian literature
Newton, Alfred, 62
Newton's gravitational theory, 211–13
New Zealand colonization, 147–54, 158nn44–45, 226–30, 232

Index

Niemann, Michelle, 104
Nil Darpan: The Indigo Planting Mirror (translation), 27, 39nn27,38
Nixon, Rob, 2, 39n33, 92, 163
nonhuman agency. *See* shared agency and consciousness
Nordhaus, Ted, 154
North and South (Gaskell), 10, 67–71, 76
The North-Star (newspaper), 205–6, 210
Nostromo (Conrad), 10, 73–78, 82n38
novels: domestic melodrama, 123–24; escapist fantasy, 241–48; human agency in, 95–96; realist novels, 21–22; systemic interconnections, as vehicle for, 121–23, 134n5. *See also* coal, as literary effect; *specific novels*

ocean, metonymy for British empire, 124–27
Omelsky, Matthew, 4, 9
On Revolution (Arendt), 77
On the Distribution of Heat over the Surface of the Globe (Hooker), 168
"On the Formation of Coal" (Huxley), 54
On the Origin of Species (Darwin), 9, 50–51, 95, 152
ontological pluralism, 41n54
Oppenheimer, Robert, 58
"Orbis Hypothesis" (Lewis & Maslin), 52–53
organicism: colonialism and, 108; defined, 104–5; origin of, 116–17n13; Ruskin's definition of life and, 11, 106–12, 115, 118n40; value and, 106–7, 113
The Oxford History of the British Empire, The Nineteenth Century, 63

pace of life, 91–93
A Pair of Blue Eyes (Hardy), 12, 187–88, 195–96
Palumbo-Liu, David, 162
Park, Geoff, 228, 229, 236
"The Passing of the Forest" (Reeves), 229, 230, 236
Peirce, Charles, 208
Pelatson, Timothy, 47–48
Perera, Suvendrini, 122
Personal Narrative of Travels to the Equinoctial Regions of the New Continent, 1799–1804 (Humboldt), 172
pessimism, 182–84, 193–94
"The Pessimism of Mr. Thomas Hardy" (Forsyth), 194

Phineas Finn (Trollope), 81n17
Physiocrats, 103–4
Pietz, William, 207
Pinkus, Karen, 12–13, 241
Pinotti, Andrea, 118n41
plays. *See* dramatic form; *specific plays*
Plotz, John, 85
Poetics (Aristotle), 23, 116–17n13
poetry. *See* elegy; *specific poems*
political economy. *See* sustainability theory
The Political Unconscious (Jameson), 65, 78–79
Poovey, Mary, 170
Post, Amy, 204
Principles of Geology (Lyell), 43, 53, 119n55
Principles of Political Economy (Malthus), 106
privatization, of water, 87–88, 96–97
Proserpina (Ruskin), 113
protectionist nationalism, 246
Purdy, Jedediah, 59

racist humor, 235f
railways, symbolism of: in *Cranford* (Gaskell), 68–69; in *Dombey and Son* (Dickens), 66, 124; in *North and South* (Gaskell), 67–68; *Rain, Steam, and Speed* (Turner), 91, 99n24
Rain, Steam, and Speed (Turner), 91, 99n24
Rajmohan's Wife (Chatterjee), 21–22
realist novels: application to environmental change, 182–84; excess and, 182–83, 189–93; hope and, 195–96; pessimism and, 182–84, 193–94; romance and, 184–89; shared agency in, 21–22, 95–96
reciprocal demand, 106
Reeves, William Pember, 228–29; "A Colonist in His Garden," 232; "The Passing of the Forest," 229, 230, 236
relational materialism, 209. *See also* Atlantic materialism
resilience. *See* hope
retrospection, 45–47, 49, 52–53
Rhodes, Cecil, 73
The Rhododendrons of the Sikkim-Himalaya (Hooker), 164, 165f
Ricardo, David, 105–6
The Rise of the Novel (Watt), 185
Robbins, Bruce, 162
rocks, vitality of, 107–8, 111
romance novels, 67–71, 76, 184–89
Romola (Eliot), 98–99n19
Ronda, Margaret, 35, 45
Rosenberg, Aaron, 11–12, 182

Ross, Kristin, 142
Rule of Darkness: British Literature and Imperialism, 1830–1914 (Brantlinger), 5–6
Ruskin, John, 101–20; "Ad Valorem," 106, 114; on aesthetic value, 113–14; on colonialism, 101–2, 107–8; on economic value, 106–7, 113; environmental criticism of, 102–3; *The Ethics of the Dust*, 103, 107–10; life, defined by, 11, 106–12, 115, 118n40; *Modern Painters*, 108; on organicism of sustainability, 103–4; *Proserpina*, 113; *The Stones of Venice*, 118n40; *The Storm-Cloud of the Nineteenth Century*, 102, 109–10; *Unto this Last*, 106; "The Work of Iron, In Nature, Art, and Policy," 103, 112. *See also* organicism; sustainability theory
Ruskin and Environment (Wheeler, ed.), 102
rusted iron, 11, 112, 115

Sabine, Edward, 131
Sade, Marquis de, 144
"Sad Joke on a Marae" (Taylor), 234
Said, Edward, 77, 82n38
Sargent, Lyman, 158n45
satire, 223–40; application to environmental change, 223–24; defined, 225; future and, 236–38; humor and, 224–30, 232–34, 236–37; irony and, 230–36; on settler colonialism, 11, 12, 141, 150, 155–56; utopian literature and, 225–26
scale: evolutionary theory and, 8–9, 183, 196; from specimen to system, 161–81; of steam power, 93; of time, 90, 94, 95–96; of water power, 88–89. *See also* botanical scale; realist novels; utopian literature
scarcity, 106, 113, 143, 243
"The Science of Fiction" (Hardy), 188–89
scientific realism, 188–89
scientific socialism, 154–55
Sconce, A., 35
sea, metonymy for British empire, 124–27
second-order thinking. *See* dialectical thinking
Seeley, J. R., 79; *The Expansion of England*, 71–73, 76
settler colonialism, 11–12, 72–73, 141–42, 147–54, 158nn44–45, 228–29, 231
shared agency and consciousness: in dramatic texts, 22–24, 32–36, 40n45; in *In Memoriam* (Tennyson), 51–52, 55–58; in *The Mill on the Floss* (Eliot), 93–94, 96–97; in realist novels, 21–22, 95–96

Shaw, David, 44–45
Shellenberger, Michael, 154
Sherburne, James Clark, 109, 110
Shewry, Teresa, 12, 223
Shuttleworth, Sally, 85–86, 90, 95, 99n22, 159n54
Sismondi, Jean Charles Léonard Simonde de, 116n2
The Sky of Our Manufacture (Taylor), 5
Slavers Throwing Overboard the Dead and Dying (Turner), 102
slavery, 74–75, 102, 204–10, 212–13
Smith, Adam, 106
Smith, Jonathan, 90, 95
socio-environmental relationships. *See* indigo cultivation and trade; Nature/Society dualism
Spear, Jeffrey, 102
species concept: defined, 61n31; in *In Memoriam* (Tennyson), 44, 47–48, 50–51, 56–57
specimens, 50–51, 163. *See also* botanical scale
spiritualism, 203–6, 211, 220n3
steam power, 75–76, 89, 91–93, 95, 98n8, 99n24. *See also* water- to steam-power transition
Steer, Philip, 10, 24, 40n53, 63, 229
Stewart, Garrett, 126–27, 132, 134n8
The Stones of Venice (Ruskin), 118n40
The Storm-Cloud of the Nineteenth Century (Ruskin), 102, 109–10
Stowe, Harriet Beecher, 204–5
strong sustainability, 105
sustainability theory: defined, 105–6; origin of, 101–3, 105–6, 114; Ruskin's definition of life and, 11, 106–12, 115, 118n40; self-contained system as, 103–4, 114–15
sustainable development, 105, 117n15
Suvin, Darko, 142, 147
Swanson, Heather, 46–47
systemic interconnections, 121–33; atmospheric imagery and, 11, 127–33; metonymy and, 123–27, 134n5; novel as vehicle for, 121–23
systems theory, 4, 10–11, 56, 141–47, 149–51, 152, 155, 183–84
Szerszynski, Bronislaw, 45

Tanoukhi, 162
Tarakratna, Ramnarayan, 27
Taussig, Michael, 25
taxidermy. *See* specimens

Taylor, Apirana: "Sad Joke on a Marae," 234
Taylor, Jesse Oak, 5, 10, 42, 121, 133
temporal structure: human perspective and, 94–95, 96; of steam power, 89, 91–93; of water power, 88–90; for water- to steam-power transition, 10–11, 85–86, 91–92, 95–97
Tennyson, Alfred, 43, 47, 49; *In Memoriam*, 42–62. See also *In Memoriam*
Tess of the D'Urbervilles (Hardy), 6–9, 13
textile industry, 88–89, 143–44
Thacker, Eugene, 194
The Theory of the Novel (Lukács), 7–8
thermodynamics, 70, 106, 121, 168
Thomas Hardy's Universe (Brenneke), 194
A Thousand Plateaus (Deleuze and Guattari), 111
The Time Machine (Wells), 142
"Time of the Icebergs" (Eggleton), 237
Tönnies, Ferdinand, 97n1
totality, 11, 123, 132, 141, 144–45, 147
Tregear, Edward, 229
Trinity Test, 58
Trollope, Anthony, 81n17; *The Fixed Period*, 149
tropes of reconciliation, 71
Tsing, Anna, 7, 46–47
Tucker, Herbert, 49
Turner, J. M. W.: *Rain, Steam, and Speed* (painting), 91, 99n24; *Slavers Throwing Overboard the Dead and Dying* (painting), 102
Two on a Tower (Hardy), 11–12, 187–88, 189–94, 198n30
typology. *See* species concept

unconformities, 48–49
Unto this Last (Ruskin), 106
Utopia (More), 225
utopian literature: as analytic tool, 154–56; application to environmental change, 139–42; colonialism depicted in, 147–54; satire and, 225–26; systems in, 142–47

value (economic), 106–7, 112–14, 115
Vashon, George, 211

Verne, Jules: *The Black Indies*, 12–13, 241–48; *The Mysterious Island*, 245–46
Vibrant Matter (Bennett), 111
Victorian Literature, Energy, and the Ecological Imagination (MacDuffie), 5, 102–3
Vidyasagar, Ishwarchandra, 28
violence: environmental, 28–29, 39n33, 227–30; of imperialism, 72–73, 74–76; of indigo cultivation, 25–26, 27–29; of modernity, 96; satire as response to, 224, 231–35; of specimen study, 50–51, 163; of water, 89, 92–93
vitalism, 105, 106–13, 115, 119n55
vital materialism, 111
Voskuil, Lynn, 11, 161

Ward, Humphrey, 197n10
water power and water rights, 86–89, 92–93, 95, 99nn21–22
water- to steam-power transition, 10–11, 85–86, 91–92, 95–97, 98n8
Watt, Ian: *The Rise of the Novel*, 185
weak sustainability, 105, 117n17
weather imagery, 11, 127–33
Weber, Max, 78
Wells, H. G.: *The Time Machine*, 142
Wenzel, Jennifer, 6
"Wessex Edition" (Hardy), 187–88
Williams, Raymond, 8, 65, 80n12, 88, 117n13; *Culture and Society*, 144
Wilson, E. O., 154
"The Wind and the Rain" (Dickens & Morley), 131
Worden, Daniel, 66
Wordsworth, William, 116–17n13
"The Work of Iron, In Nature, Art, and Policy" (Ruskin), 103, 112
World Commission on Environment and Development (WCED), 105
world-systems analysis, 41n54, 145–46, 155
Worster, Donald, 103
Wrigley, E. A., 67

Yaeger, Patricia, 80n11

Zemka, Sue, 148

www.ingramcontent.com/pod-product-compliance
Lightning Source LLC
Chambersburg PA
CBHW030437300426
44112CB00009B/1041